RENEWABLE DIESEL

RENEWABLE DIESEL

Value Chain, Sustainability, and Challenges

Edited by

DIPESH KUMAR

Nalanda University Centre Neemrana, Tribhuvan College of Environment and Development Sciences, Rajasthan, India

BHASKAR SINGH

Department of Environmental Science, Central University of Jharkhand, Ranchi, India

SANJAY KUMAR GUPTA

Environmental Engineering, Department of Civil engineering, Indian Institute of technology Delhi, Hauz Khas, New Delhi, India

ELSEVIER

Elsevier
Radarweg 29, PO Box 211, 1000 AE Amsterdam, Netherlands
The Boulevard, Langford Lane, Kidlington, Oxford OX5 1GB, United Kingdom
50 Hampshire Street, 5th Floor, Cambridge, MA 02139, United States

ISBN: 978-0-323-91153-5

For Information on all Elsevier publications
visit our website at https://www.elsevier.com/books-and-journals

Publisher: Megan R. Ball
Acquisitions Editor: Peter Adamson
Editorial Project Manager: Aleksandra Packowska
Production Project Manager: Surya Narayanan Jayachandran
Cover Designer: Vicky Pearson Esser

Typeset by MPS Limited, Chennai, India

Contents

List of contributors

Adewale Adewuyi Department of Chemical Sciences, Faculty of Natural Sciences, Redeemer's University, Ede, Osun State, Nigeria

Abhishek Anand Department of Chemical Engineering, Birla Institute of Technology and Science (BITS), Pilani, Rajasthan, India

Ishita Banerjee Department of Biochemistry & Cellular and Molecular Biology, The University of Tennessee, Knoxville, TN, United States

Rachael J. Barla Department of Chemical Engineering, Birla Institute of Technology and Science (BITS), Pilani, Rajasthan, India

Mahnoush Beygisangchin Institute of Nanoscience and Nanotechnology, Universiti Putra Malaysia, Serdang, Selangor, Malaysia

Chandrama Roy Chowdhury Department of Food Science, Czech University of Life Sciences, Prague, Czech Republic

Arpita Das Department of Chemistry, National Institute of Technology Silchar, Silchar, Assam, India

Swapnamoy Dutta Bredesen Center for Interdisciplinary Research and Graduate Education, the University of Tennessee, Knoxville, TN, United States

Safa Gamal Putra Laboratory for Sustainable Bioenergy and Chemicals, Faculty of Science, Universiti Putra Malaysia, Serdang, Selangor, Malaysia; Department of Chemistry, Faculty of Science, Universiti Putra Malaysia, Serdang, Selangor, Malaysia

Suresh Gupta Department of Chemical Engineering, Birla Institute of Technology and Science (BITS), Pilani, Rajasthan, India

Balkis Hazmi Institute of Nanoscience and Nanotechnology, Universiti Putra Malaysia, Serdang, Selangor, Malaysia

Km Smriti Jaiswal Department of Chemistry, Deen Dayal Upadhyay Gorakhpur University, Gorakhpur, Uttar Pradesh, India

Krishna Kumar Jaiswal Department of Green Energy Technology, Pondicherry University, Puducherry, India

Gaurav Kant Department of Biotechnology, Motilal Nehru National Institute of Technology Allahabad, Prayagraj, Uttar Pradesh, India

Chuan Li Lee Institute of Tropical Forestry and Forest Product, Universiti Putra Malaysia, Serdang, Selangor, Malaysia

H.M.D. Nisansala Nanocomposite Research Centre, Faculty of Applied Sciences, University of Sri Jayewardenepura, Nugegoda, Western Province, Sri Lanka; Faculty of Graduate Studies, University of Sri Jayewardenepura, Nugegoda, Western Province, Sri Lanka

Ashutosh Pandey Department of Biotechnology, Motilal Nehru National Institute of Technology Allahabad, Prayagraj, Uttar Pradesh, India; Institute for Water and Wastewater Technology, Durban University of Technology, Durban, South Africa; Faculty of Life Science and Technology, Department of Biotechnology, AKS University, Satna, Madhya Pradesh, India

Dipesh Shikchand Patle Department of Chemical Engineering, Motilal Nehru National Institute of Technology Allahabad, Prayagraj, Uttar Pradesh, India

Smita Raghuvanshi Department of Chemical Engineering, Birla Institute of Technology and Science (BITS), Pilani, Rajasthan, India

Umer Rashid Institute of Nanoscience and Nanotechnology, Universiti Putra Malaysia, Serdang, Selangor, Malaysia; Center of Excellence in Catalysis for Bioenergy and Renewable Chemical (CBRC), Faculty of Science, Chulalongkorn University, Bangkok, Thailand

Samuel Lalthazuala Rokhum Department of Chemistry, National Institute of Technology Silchar, Silchar, Assam, India

B. Sangmesh Department of Green Energy Technology, Pondicherry University, Puducherry, India

Veena Singh Department of Material Science and Engineering, KTH-Royal Institute of Technology, Stockholm, Sweden

N.M.S. Sirimuthu Nanocomposite Research Centre, Faculty of Applied Sciences, University of Sri Jayewardenepura, Nugegoda, Western Province, Sri Lanka

Sameer Srivastava Department of Biotechnology, Motilal Nehru National Institute of Technology Allahabad, Prayagraj, Uttar Pradesh, India

Preface

The world is currently facing a critical dependence on petroleum, often imported. This dependence has resulted in environmental pollution and climate change, among the prominent drivers for promoting and developing alternative energy/fuels. As a result, there is a massive transition towards alternative and renewable forms of energy. The electricity sector is leading, but other hard-to-abate sectors still await a techno-economically viable and environmentally appealing alternative. The transport sector, for instance, is still facing some challenges despite witnessing a significant surge in the demand and availability of battery-powered powertrains. Limited availability of metals used in batteries, high cost, range anxiety, high weight of batteries, inadequate availability of charging stations, unsuitability for long haul and heavy-duty vehicles and lack of end-of-life management policies are some challenges. Even though the numbers are increasing, e-vehicles still constitute less than 5% of the total vehicular population on the road, and the transition to e-mobility is still not a global phenomenon. Any significant substitution of internal-combustion engine-based vehicles is only expected in the mid to long term. However, the climate cannot wait, and we must continue to work toward finding more environment-friendly alternatives.

An alternative arrangement that is compatible with the existing infrastructure of internal-combustion engines is inherently attractive as changing over to an entirely new fleet of powertrains is daunting and can only be justified in the mid to long run. One option that has been around for some time is alternative fuels derived from oleaginous biomass. Different forms of vegetable oil, such as virgin oil and those derived through blending, pyrolysis, emulsification and transesterification of oleaginous biomass, have all been explored and used. The most commonly produced and used fuel among these is biodiesel, which, although possessing advantages over conventional mineral diesel, also comes with its own set of limitations and challenges. These include its limited miscibility in diesel, poor atomisation and cold-flow properties and inferior stability. The incompatibility of higher biodiesel blends in compression-ignition engines is a significant deterrent for the industry. However, an alternative thermo-chemical treatment of oleaginous biomass, popularly known as hydro-processing or hydro-treatment, is known to overcome these limitations and yields a drop-in fuel called "renewable diesel (RD)." The chemical composition of RD is similar to that of petroleum diesel, and it can utilise the existing infrastructure of refineries, distribution systems and powertrains. It is clear that RD is an up-and-coming alternative worth considering.

Interestingly, despite RD's potential and growing appeal, the existing literature on the subject is relatively scanty and scattered. Different terms are often used to refer to the same fuel. Green diesel and hydro-treated/hydro-processed vegetable, second-generation biodiesel oil, etc. are various other terms used to refer to RD, considering the biological origin of the feedstock (free fatty acids and triacylglycerol), which is typical for almost all of the biofuels. Terminology is essential to define the subjects of discussion, much like how it is crucial in chemistry.

This edited volume titled *Renewable Diesel: Value Chain, Sustainability, and Challenges* by Dipesh Kumar, Bhaskar Singh and Sanjay Kumar Gupta is fascinating and informative. The book aims to compile some critical elements in the value chain and provides a thorough overview of RD. It is fascinating how this volume distinguishes between alternative fuels derived from oleaginous biomass. The volume covers literature on the synthesis of RD (Chapter 1), its quality specifications (Chapter 2), life cycle–based appraisal (Chapter 3), challenges and opportunities (Chapter 4) and the current market status and future industry outlook (Chapter 5). The book also presents a comparative analysis of biodiesel and RD in Chapter 6. Finally, the book also includes a chapter on fuels derived from waste plastics to supplement the research and developmental efforts aimed at developing other alternative fuels (Chapter 7). This volume is an excellent source of well-researched information for scholars keen to learn about alternative fuels.

This edited volume provides a wealth of pragmatic studies and reviews of contemporary developments and opportunities in the field of RD.

Dipesh Kumar

1

Synthesis of renewable diesel as a substitute for fossil fuels

Krishna Kumar Jaiswal[1], Chandrama Roy Chowdhury[2], Swapnamoy Dutta[3], Ishita Banerjee[4], Km Smriti Jaiswal[5], H.M.D. Nisansala[6,7], B. Sangmesh[1] and N.M.S. Sirimuthu[6]

[1]Department of Green Energy Technology, Pondicherry University, Puducherry, India [2]Department of Food Science, Czech University of Life Sciences, Prague, Czech Republic [3]Bredesen Center for Interdisciplinary Research and Graduate Education, the University of Tennessee, Knoxville, TN, United States [4]Department of Biochemistry & Cellular and Molecular Biology, The University of Tennessee, Knoxville, TN, United States [5]Department of Chemistry, Deen Dayal Upadhyay Gorakhpur University, Gorakhpur, Uttar Pradesh, India [6]Nanocomposite Research Centre, Faculty of Applied Sciences, University of Sri Jayewardenepura, Nugegoda, Western Province, Sri Lanka [7]Faculty of Graduate Studies, University of Sri Jayewardenepura, Nugegoda, Western Province, Sri Lanka

1.1 Introduction

Energy resource is one of the essential needs in the present-day scenario of world today. The extensive use of nonrenewable energy has posed a threat to the global environment as it contributes to global warming and pollution. Also, the inadequate availability of fossil-based fuels is another persistent confront. The limited supply and availability of nonrenewable resources have increased the price of fuel energy, putting financial stress on most people (Shamoon et al., 2022). It also plays an increasing role in hampering and degrading climate conditions around the world. The use of fossil fuels in the automotive industries generates many exhaust gases known as greenhouse gases. The greenhouse gases increase the carbon footprint and contribute to global warming statuses and climate changes. These environmental threats have determined the exploration of alternative energy sources (renewable energy) to reduce the disadvantages of nonrenewable fuels (Heidari & Pearce, 2016). Researchers are now focusing on making renewable fuels and biofuels from biomass and cellulosic materials because the conversion process is more straightforward and affordable.

Renewable Diesel. DOI: https://doi.org/10.1016/B978-0-323-91153-5.00001-7

Biofuels are the ecofriendly alternative fuels or energy resources generated from bio-based material and biomass. It is generally catagorized into the 1st generation, 2nd generation, and 3rd generation biofuels (Jaiswal, Banerjee, & Mayookha, 2021; Jaiswal, Dutta, Banerjee, Pohrmen, & Kumar, 2021; Jaiswal, Kumar, Vlaskin, & Nanda, 2021). 1st generation biofuels include the biofuels which is directly related to the renewable biomass that is more than often edible resources such as biodiesel, bioethanol, etc. 2nd generation biofuels are well-defined as biofuels generated from a wide-ranging feedstocks, mainly but not restricted to nonedible lignocellulosic biomass. The most established explanation for 3rd generation biofuels is a renewable biofuels generated from microalgal biomass, with a prominent growth and yield in comparision to the typical lignocellulosic biomass (Alalwan, Alminshid, & Aljaafari, 2019; Arora et al., 2020; Jaiswal & Prasath, 2016). Recently, the production of bioethanol from the corn and other grains feedstock has been recorded at approximately 15 billion gallons per year. It can also be said that the production of biofuel from crops is equivalent to 10% of current gasoline consumption. Some prominent examples of biofuels can be bioethanol (produced mainly from sugar cane or corn), biogas (generated from vegetable waste, animal manure or other organic materials) and biodiesel (made from animal and vegetable fats/lipids) (Choudhary, Joshi, Rao, & Srivastava, 2021; Jaiswal, Banerjee et al., 2021; Jaiswal, Dutta et al., 2021; Jaiswal, Kumar et al., 2021).

Biodiesel produced from lipids extracted from biomass mainly comprises fatty acid methyl esters (FAMEs). It also contains a small amount of nonester fuel. The use of biomass for generating and developing a renewable energy source is suitable for its neutrality against CO_2. Researchers have also focused on producing biofuels from agricultural residues, as they are also good sources of lignocellulose (Limayem & Ricke, 2012). Some of the significant advantages of biomass-based biofuels include the reduced generation of greenhouse gases such as methane, carbon dioxide and nitrous oxide. It also releases fewer pollutants. Therefore it can be said that they contribute less to global warming. They are highly renewable, profitable and efficient. They have a minimal carbon footprint, making them suitable for environmental health and safety. It has been proven that biodiesel can reduce 47% of particles in the atmosphere and 67% of hydrocarbon emissions and reduces smog (smoke + fog). Using agricultural waste in biofuel production can also help to reduce waste problems by 96%. This, in turn, contributes to a 79% reduction in wastewater generation (Ambaye et al., 2021). However, one of the most significant disadvantages of biofuel production is that they are time-consuming and laborious to develop. The cost of production also depends on the complexity of the production process.

1.2 Biodiesel and renewable diesels

Biodiesel is a category of fuel that presents itself as an alternative candidate to the predominant petro-diesel fuels. This type of disel fuels is generally composed of long-chains of methyl or ethyl fatty esters produced from biomass resources that maintain specific standards such as ASTM D6751, EN 14214 (Upare & Upare, 2020). In the early 1930s, it was first observed and soon recognised as a potential substitute for conventionally used transportation fuels. Biodiesel is a fascinating option, and it has been widely explored in recent decades due to specific advantages such as easy-to-obtain feedstocks (e.g., crude vegetable oils, and animal fats, etc.) for transesterification (as an easy conversion technique), biodegradability, less toxic impact on nature, etc. Vegetable oils derived from the seeds of the crops (generally from food and nonfood crops such as rapeseed, sunflower, soybean, switchgrass and jatropha) are used to convert them into methyl esters of fatty acids or precisely into biodiesel through a transesterification reaction (Folayan, Anawe, Aladejare, & Ayeni, 2019). However, in later times, it was discovered that the exploitation of energy crops to generate biodiesel was hampering economic stability and posing food security problems. Therefore crop residues such as leaves, stems, hulls, municipal and industrial waste, waste animal fats/lipids and microalgae, were seriously considered for biodiesel feedstock (Jaiswal, Banerjee, et al., 2021; Jaiswal, Dutta, et al., 2021; Jaiswal, Kumar, et al., 2021; Jaiswal & Pandey, 2014; Kumar et al., 2020; Kumar, Jaiswal, et al., 2021; Kumar, Yogeshwar, et al., 2021).

Aside from biodiesel, renewable diesel is another vital form of fuels that can be applied effectively as a fuel for automobiles. It has been manufactured maintaining the following standards: ASTM D975 and EN 590. It contains mainly long-, short- and branched-chain alkanes and negligible aromatic compounds. Similar triacylglyceride feedstocks involved in biodiesel production (especially vegetable oil) can also be utilized to generate renewable diesels by the catalytic hydrogenation method using the temperature range of 300°C–400°C together with pressures of ~ 100 atm. The hydrogenation or hydrotreating process involves the generation of hydrocarbons in the C15–C18 range and eradicates oxygen, nitrogen and sulphur. Unlike biodiesel, renewable diesel has superior cleaner combustion and better cold flow characteristics (Singh, Subramanian, & Garg, 2018). A deeper analysis of renewable diesel showed few similar and excellent properties compared to petro-diesel. Still, it has numerous differences compared to biodiesel, elaborated in other sections.

1.3 Renewable diesel

Renewable diesel is the type of fuel generally obtained from various vegetable oils and possesses the properties analogous to petro-diesel fuels. Previously, the phrase 'renewable diesel' was used to identify the type of fuel produced from a renewable source without specifying any reference to the characteristic of the disel fuels generated. On the other hand, the phrase 'green diesel' considers the origin of the fuel and the impact on greenhouse gases due to its particular use. So the expressions 'green diesel' and 'renewable diesel' are much more complex (Bezergianni & Dimitriadis, 2013). In many cases, the interchangeable use of these terms was not entirely clear to make a precise distinction. However, recent studies have clarified the gap in understanding and have suggested their complementary use of the term (green diesel or renewable diesel) and, in addition, have labelled them as second-generation diesel or bio-hydrogenated diesel. Vegetable oils and fats as renewable sources are commonly used with catalytic hydroprocessing to produce renewable diesel. It should not be confused with biodiesel since most similar oils and fats are used to synthesise biodiesel through the transesterification reaction. Although, FAME or biodiesel formed from vegetable oils and/or animal fats can be utilized to generate low-range hydrocarbons and fuels that can easily be compared as almost equivalent to renewable diesel production. Using processes such as catalytic hydrodeoxygenation (HDO) or hydroprocessing reaction (along with steam cracking), the upgrading of FAME biodiesel derived from rapeseed or residual cooking oil to diesel as alkanes (C15–C18) or suitable hydrocarbons (C2–C4) have already been reported with optimised reaction parameters (Pattanaik & Misra, 2017).

Another way to produce renewable diesel from biomass feedstock is through the Fischer–Tropsch (FT) process. The FT process can effectively realise the syngas gas for further conversion of higher hydrocarbons. Furthermore, the beneficial factor of this process is that it can use a variety of feedstocks and constantly maintains the quality of the fuel. Mainly, this process was used to produce wood, natural gas and coal to produce hydrocarbons. Still, waste from the paper industry, cellulose and agricultural/forestry residues showed promising results in later times. The FT procedure to convert biomass into renewable diesel involves the following series of processes: syngas generation, catalytic conversion of syngas and finally, hydroprocessing (Douvartzides, Charisiou, Papageridis, & Goula, 2019). In addition, the FT process is classified according to the temperature range, where the low-temperature FT process includes a temperature range of 180°C–250°C with the use of cobalt catalysts favourably. Besides, the high-temperature process is considered with a range of temperature to be 330°C–350°C together with iron-based catalysts. These renewable diesels derived from the FT process are composed of low aromatic compounds. As a result, cleaner combustion can be achieved, specifically, reduced NO_x emissions, making it better fuel for commercial use (Smagala et al., 2013).

'White diesel' is another variety of renewable fuel that contains paraffinic fuel without traces of aromatic compounds and is mainly produced from waste cooking oil. This type of diesel has exhibited optimum heating value, high cetane number (CN), negligible sulphur levels (1.54 ppm) and superior oxidation stability but has problems with its cold flow properties. Catalytic hydrotreating and hydrotreating together with isomerisation are two processes used to prepare white diesel. It is currently produced on a large scale in the pilot hydroprocessing units of the Center for Research and Technology, Hellas, Greece. More importantly, the conversion efficiency of used cooking oil (UCO) into white diesel is almost 92% (Bezergianni & Dimitriadis, 2013).

Besides these different types of renewable diesel, hybrid biodiesel prepared by co-processing petroleum oil with vegetable oils (soybean, rapeseed, jatropha, coconut and palm oil) is also considered a type of renewable diesel. The hydroprocessing of petroleum fuel with vegetable oil is known as H-Bio. 'Hybrid diesel' can also be formed by blending renewable diesel with petro-diesel without any technical alterations to diesel engines. The performance is almost similar to that of individual engines. Hybrid diesel can be effective as many countries are fully or partially dependent on petroleum fuel. Still, the technology is under development, and hopefully, it may lead to a potential breakthrough in the coming days. In this way, we can differentiate different types of renewable fuels (Amin, 2019). However, renewable diesel has not been categorised typically, but based on the nature and properties of these almost similar fuels, we have termed these fuels as renewable fuels.

1.4 The feedstock of renewable diesel

To substitute the widespread use of fossil-based fuels, it became necessary to establish and expand the operation of nature-friendly biofuels for commercial use. Therefore to make these biofuels more accessible, cheap and readily available, feedstocks, including edible and nonedible sources, are rapidly exploited for commercial purposes. In particular, vegetable oils and animal fats are considered the best materials for the synthesis of biodiesel and renewable diesel using optimal reaction conditions and various catalysts. Compared to biodiesel, renewable diesel has a much wider range of compatible feedstocks; since renewable diesel is generated over the saturation of triglycerides by means of hydrogen and produces mixtures of saturated hydrocarbons regardless of the feedstocks used initially, while during the production of biodiesel, a blend of unsaturated FAMEs have been formed when the unsaturated feedstock is used. Therefore it can create oxidative instability (Pelemo, Inambao, & Onuh, 2020).

Studies have also confirmed that algae are readily available and inexpensive biomass to produce these diesel fuels. In addition, the ability to grow in nonarable places, the substantial growth rate and recent genetic manipulation to improve the ability of algal biomass are more fascinating. The oils obtained from soybean, palm, sunflower seed, jatropha seed, rubber seed, rapeseed, castor seed, and waste cooking oil, etc., are usually used as feedstocks to produce renewable diesel (Pelemo et al., 2020). Although in recent times, edible feedstocks have been chiefly avoided to produce fuel, as it can present a potential risk of limiting the food supply and, as a result, the socio-economic balance can be compromised. Due to their abundance, lignocellulosic biomass materials are considered one of the leading biomass feedstock options. Waste from forestry, agriculture and food are viable options. Using it to produce fuel to contribute to the energy sector can be a breakthrough in waste disposal management techniques. In this sense, rice straw, wheat straw, paper pulp/sludge, defatted rice bran, orange peel, olive pruning, animal/fish fats, etc., have been investigated to verify their feasibility in producing renewable biodiesel (Jaiswal, Dutta, Banerjee, Mayookha, & Bhushan, 2022; Jaiswal, Jha, & Prasath, 2014). It is imperative to identify possible sources of waste that can be used to produce renewable diesel, as it will not encourage advances in future energy generation. Still, it will also help provide effective waste management policies (Patel & Kumar, 2016). Therefore to select biomass materials to generate renewable diesel, the essential factors to consider are the constant supply of feedstocks, the maintenance of a prosperous economy, the balance of the carbon footprint of these produced renewable diesel fuels, etc. In addition, the evaluation of the compositions of fatty acids composition of the used oil feedstocks and the structural analysis (cellulose, hemicellulose, lignin, etc.) of the biomass feedstocks intended to be used to generate renewable diesels will help to provide a clear view of the conversion and estimate its feasibility for real-world application.

1.5 Production of biodiesel

Biodiesel fuels have been generated via the transesterification conversion of lipid-containing feedstocks, including edible or non-edible oils/fats, cyanobacteria or microalgae. Mainly, edible oil feedstocks exhibited higher conversion efficiency, but due to socio-economic reasons, nonedible oils such as jatropha and castor oil were considered valuable candidates. Along with the conversion yield, selecting the optimal feedstock is necessary to produce biodiesel with precise composition and reasonable price (Jaiswal, Kumar, Vlaskin, & Nanda, 2020; Kim, Sen, & Maravelias, 2013).

Transesterification is the fundamental procedure for generating FAMEs or biodiesel. Feedstocks composed of oils or fats with specially selected alcohols and various catalysts can be used (Fig. 1.1). The reaction converts triglyceride molecules (in oil or fat) to FAME. The triglyceride units are broken through

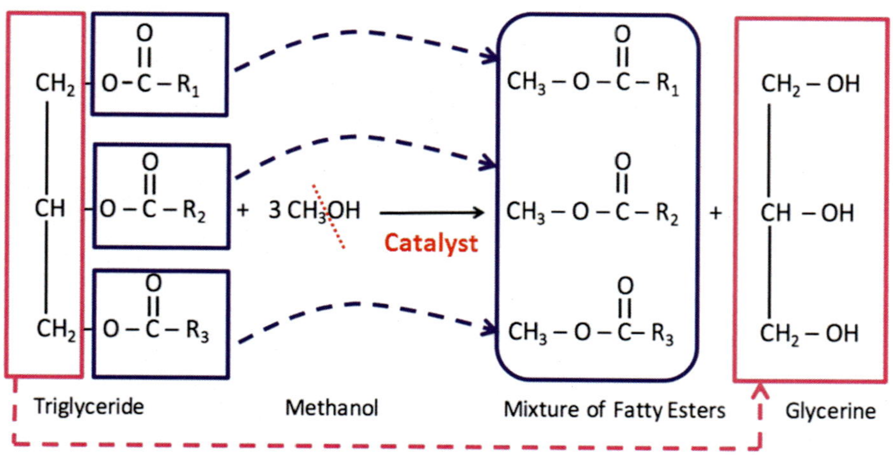

Figure 1.1 General mechanism of transesterification of lipid triglycerides to produce biodiesel.

the following steps: forming the tetrahedral intermediate, the subsequent generation of fatty acid esters, diglyceride ions from the unstable intermediate tetrahedral dissociation and then through proton transfer, recovery of the catalyst. Finally, three fatty acid ester units and one glycerol unit have been produced. Experimental parameters such as the nature and amount of catalysts, oil/alcohol ratio, temperature and reaction duration affect biodiesel formation in terms of quality and quantity (Kiss & Bildea, 2012). Based on the nature of the catalysts, the transesterification reaction can be classified into mainly three divisions, i.e., homogeneous, heterogeneous and enzyme-catalysed transesterification processes. Furthermore, they are subcategorised according to the acidic or alkaline nature of the catalysts.

The intermediate steps of biodiesel synthesis involving the different catalysts mentioned above varied in their favourable reaction conditions with yield efficiency. However, this conventional transesterification reaction has presented a few pitfalls, and improved techniques have been developed to combat them due to the use of methanol and oil as immiscible reagents. To avoid this issue, ultrasonic irradiation-guided transesterification has been proposed in studies where sound waves of suitable frequency (24–40 kHz) have been applied to the reaction mixture to penetrate the phase boundary between methanol and oil to form a well-mixed emulsion blend finally. The reaction time, catalyst concentration, and amount of alcohol need to be much lesser than conventional transesterification (Bhatia et al., 2021). Biodiesel production using different feedstocks and catalysts under various reaction parameters is represented in Table 1.1. The supercritical alcohol process is another superior transesterification reaction in which the pressure (80 atm) and temperature (300°C–400°C) have been set above the critical level of pressure and temperature. The supercritical alcohol method shows a faster reaction rate and better mass transfer, like the ultrasonic irradiation technique. In addition, the supercritical alcohol process is largely catalyst-free, which reduces overhead expansions and avoids the complexity of separating biodiesel produced with catalysts (Tabatabaei et al., 2019).

1.6 Production of renewable diesel

Similar to biodiesel, the production of renewable diesels can be done using a wide range of vegetable and waste oils. In addition, feedstocks such as fish lipids and animal fats can also be utilized. The establishment of techniques to process leftovers

Table 1.1 Biodiesel production using different feedstocks and catalysts under various reaction parameters.

Lipid feedstocks	Catalyst	Concentration of catalyst (wt.%)	Alcohol:lipid (molar ratio)	Time (min)	Temperature (°C)	Yield (%)	References
Palm oil	KOH	1.0	8:1	60	60	88	Ali and Tay (2013)
Tallow seed oil	KF/CaO	4.0	12:1	150	65	96.8	Wen, Wang, Lu, Hu, and Han (2010)
Soybean oil	CaO/*n*-Al_2O_3	3.0	9:1	600	150	90	Pasupulety, Gunda, Liu, Rempel, and Ng (2013)
Castor oil	Fe–ZnO	14.0	12:1	50	55	91	Baskar and Soumiya (2016)
Canola oil	K_2CO_3/Al_2O_3	3.16	11.48:1	120	60	96.3	D'Cruz, Kulkarni, Meher, and Dalai (2007)
Waste cooking oil	H_2SO_4	2.0	10:1	240	95	>90	Wang, Ou, Liu, Xue, and Tang (2006)
Used frying oil	NaOH	1.1	7:1	20	60	88.8	Leung and Guo (2006)
Mutton fat	KOH/MgO	4.0	22:1	20	65	>98	Mutreja, Singh, and Ali (2011)
Fish lipid	ZnO	2.5	—	240	70	89.86	Dutta, Jaiswal, Verma, Basavaraju, and Ramaswamy (2019)

from the food industry or waste materials to produce renewable diesel will allow solving both the energy supply problems and management of leftover waste. Although biodiesel and renewable diesel are made from similar feedstocks, their composition varies from each other since the preparation process is quite diverse. Renewable diesels are mainly consisted of straight- and branched-chain C15–C18 paraffins. HDO is the critical process of producing renewable diesel using the aforementioned feedstocks (Bezergianni & Dimitriadis, 2013). The triglyceride molecule has been fragmented into three molecules by eliminating oxygen during interaction with hydrogen molecules throughout the reaction period. As an outcome of the abolition of oxygen molecules in the reactions, hydrocarbon molecules analogous to the components present in diesel are generated. However, different technologies are also being developed to create renewable diesel. Based on other feedstocks or conversion routes, we have already discussed different types of renewable diesel in previous sections, which we have illustrated here for better understanding.

Generally, catalytic hydroprocessing is a technique used in petroleum refineries to improve the hydrogen/carbon ratio, reduce the boiling point of refinery products and control the concentration of heteroatoms and metals. This technique can also be used to produce renewable diesel through vegetable oils/fats that can be applied in existing diesel engines with lesser environmental impact. This technique, as mentioned, mainly involves two main steps: a catalytic hydrotreating reaction phase to produce normal paraffin and thereafter a catalytic isomerisation phase which results in a mixture of *n*- and isoparaffin. Apart from vegetable oils/fats, biodiesel generated from oils and fats can also be converted into renewable diesel (containing low-range hydrocarbons) by applying catalytic hydroprocessing or upgrading (Speight, 2004). Biodiesel produced through transesterification from oils and fats is further hydroprocessed or HDO to prepare renewable diesel. The production of renewable diesels in the attendance of different catalysts and reaction parameters has been shown in Table 1.2. The prepared diesel exhibited better kinematic viscosity and specific gravity and improved low-temperature operation/performance and oxidation stability.

The FT process is another effective method of producing renewable diesel. Previously, this method was applied to derive hydrocarbons from coal. Later, using biomass and natural gas, this method effectively harvested fuels and lubricated oils. Configurations of sludge or fixed bed reactors have been developed to convert different types of biomass to generate

Table 1.2 Renewable diesel production in the presence of different catalysts and reaction parameters.

Feedstock	Catalyst	Liquid hourly space velocity (h^{-1})	Pressure (Bar)	Temperature (°C)	Yield (%)	References
Sunflower oils	$Pt/HZSM\text{-}22/Al_2O_3$	2.0–3.0	50–60	320–330	94–95	Hancsók et al. (2007)
Jatropha curcas oil	Ni–Co/MWCNTs	—	0.01	350	80	Asikin-Mijan, Lee, Abdulkareem-Alsultan, Afandi, and Taufiq-Yap (2017)
Microalgae oil	Pt/Al_2O_3	—	33.4	310	76.5	Zhou and Lawal (2016)
Waste chicken fats	$Ni/\gamma\text{-}Al_2O_3$	0.5	50	330	82	Kaewmeesri, Srifa, Itthibenchapong, and Faungnawakij (2015)
Canola oil	$Ni\text{–}Mo/\gamma\text{-}Al_2O_3$	1.0	31.02	375	100	Taromi and Kaliaguine (2018)
Microalgae oil	$Ni\text{–}Mo/\gamma\text{-}Al_2O_3$	5	34.47	360	98.7	Zhou and Lawal (2015)
Sunflower oil	NiMo	5.2	50	350	95	Huber, O'Connor, and Corma (2007)
Beechwood	Ru/C	—	200	350	90	Wildschut, Mahfud, Venderbosch, and Heeres (2009)
Crude palm oil	$Ni\text{–}Mo/Al_2O_3$	—	55	315	68.2	Zikri and Aznury (2020)
Microalgae oil	Ni/ZrO_2	—	40	260	70–75	Peng, Yuan, Zhao, and Lercher (2012)

productive elements through the FT process. Syngas generation, syngas catalytic conversion and finally, hydroprocessing are the three vital steps of the FT process. Renewable diesel produced through the FT process or FT diesels have a greater calorific value (43–45 MJ/kg) and low levels of sulphur and nitrogen compared to fossil fuels (Madden, Alles, & Demirel, 2021).

On the other hand, white diesel produced from UCO is another product similar to the type of renewable diesel discussed earlier. The development of white diesel mainly includes three primary reactions: decarbonylation, decarboxylation and deoxygenation. Firstly, cooking oil and hydrogen are mixed in a 1:500 volume ratio in a hydroprocessing unit, and, in the presence of catalysts, the reaction takes place. Subsequently, the liquid and gaseous products formed are separated, where the liquid products contain paraffin and isoparaffins together with water, where the water is carefully precipitated and isolated. Finally, the remaining liquid product has been fractionated and subjected to fractionation to prepare the resulting fuel, called 'white diesel' (Chrysikou, Dagonikou, Dimitriadis, & Bezergianni, 2019). This type of renewable diesel has almost negligible oxygen content and no heavy molecules. In this way, as we detail in this section, different renewable diesel can be prepared.

1.7 Effect of operating parameters for production of renewable diesel

In order to shrink carbon dioxide emission and encounter the growing energy mandate, various studies have been performed to produce renewable fuels from sustainable, profitable and environment-friendly biomass. This renewable fuel can be utilized instead of petroleum-based fuels (Long et al., 2021). This third-generation renewable diesel is driven by catalytic HDO of vegetable oil. Renewable diesel contains a low amount of aromatic compounds and shorter-chain fatty acids. This can be used in place of petroleum fuel due to some of its superior qualities, such as cleaner burning and better cold flow properties. Compared with biodiesel (FAMEs), renewable diesel has higher oxidation stability, lower specific gravity and higher CN. When blended with petroleum diesel, it has better cold flow properties. In addition, renewable diesel is compatible with petro-diesel engines, hence no modifications are needed. Renewable diesel does not contain oxygen, so the oxidation stability is higher, noncorrosive and possesses higher heating value similar to petro-diesel. Various hydrocatalytic technologies have been studied and used to generate renewable biodiesel from bio-oil (Ren et al., 2020). Generally, renewable diesels can be generated from biomass via mainly four technologies: hydroprocessing; catalytic upgrading of sugars, starches and alcohols; thermal (pyrolysis) conversion and bio-oil upgrading; and biomass-to-liquid thermochemical approaches. It is significant to note that the factors that affect the hydrotreating technique are temperature, hydrogen/oil ratio, pressure, catalysts, and space velocity (Niu et al., 2020). The main factors that need to be considered to produce renewable diesels are the nature of feedstocks and the choice of catalyst.

1.7.1 Feedstock to produce renewable diesels

The choice of feedstock can significantly influence renewable diesel production through the hydrogenation process. Edible and inedible oils derived from plants such as rapeseed, soybean, palm, jojoba, jatropha, etc. can be used as the feedstocks (Shaah et al., 2021). Animal fat and microalgae lipids are other important sources of renewable biodiesel production. UCOs are the inexpensive and second most abundant feedstock source for renewable biodiesel production. UCO is waste oil produced from vegetable oils after frying and is abundant and readily obtainable in food processing industries, restaurants, households and fast food outlets. The properties of the fatty acids of UCO make it the perfect feedstock to produce renewable diesels. Vegetable oil used for frying endures a number of physical and chemical fluctuations. During frying, certain undesirable complexes are formed, such as free fatty acids and polymerised triglycerides, which cause an increase in molecular weight and condense the volatility of the oils. UCO is a renewable source and does not comprise aromatic, metallic or sulphurous impurities and high acidity values. Therefore reusing this UCO for renewable diesel production actively reduces the chances of health problems, such as hypertension, diabetes and vascular inflammation, from consuming UCO after a single use (de Onís, 2021). Along with UCO, sugars and starches, the primary feedstock for ethanol production, they are excellent feedstock options for renewable diesel production by converting higher hydrocarbons by catalytic upgrading (Tuli & Kasture, 2022). Herbaceous biomass and woody biomass contain lignocellulose suitable for thermal or thermochemical conversion, and finally, renewable diesel.

1.7.2 Catalysts to produce renewable diesels

The choice of catalysts is critical to improve the qualitative and quantitative yields of renewable diesel production. The use of catalysts for producing renewable biodiesel is to break bonds, and these it is very likely to influence the route and degree of breaking of the C—C and C—O bonds (Abdulkareem-Alsultan et al., 2019). The commonly used hydroprocessing catalysts are Ni, Co and Mo as active metals which are generally reinforced by alumina or silica-alumina. In addition, Co–Mo and Ni–Mo catalysts with alumina (γ-Al_2O_3) have been extensively applied for hydrogenation, HDO and hydrocarboxylation, which are also suitable desulphurisation catalysts.

1.7.2.1 Traditional catalysts

Several studies have reported that sulphurised Co–Mo/γ-Al_2O_3 and Ni–Mo/γ-Al_2O_3 are the active catalysts for hydrodesulphurisation (HDS) that are mainly applied in sulphur elimination in renewable diesel production. However, the choice of catalysts usually depends on the production routes, e.g., combined Ni–Mo/sulphurised Al_2O_3 catalysts provide better performance for the diesel range of C15–C18 products from a mixture of residual soybean oil and gas oil due to nominal cracking. In contrast, a sulphurised Ni–W/SiO_2–Al_2O_3 catalysts provide a significant jet range product due to catalytic cracking (Shirasaki, Nasu, Hashimoto, & Ishihara, 2019). ZSM-5 with Ni/Mo or Ni/W catalysts has been used to produce jet biofuel using jatropha oil as feedstock through a single-step route. The selection of the route and nature of catalyst used can determine the cost and viability of resultant optimal yield of renewable diesel.

1.7.2.2 Noble metal catalysts

Along with traditional catalysts, studies have shown that the performance of noble metals as a catalyst is better than conventional refinery catalysts in terms of yield. Among the noble metals, Pd exhibited the premier activity, followed by Rh, with Pt representing the lower activity due to incomplete reduction at the time of reaction. Noble metals can be easily poisoned by small amounts of sulphur compounds existing in bio-oils, which are usually costly (Bjelić, Grilc, & Likozar, 2020).

1.7.2.3 Transition metal catalysts

Some inexpensive transition metals can be used as catalysts for hydroprocessing in different reactor systems. Ni–Cu/TiO_2 catalysts showed the best hydrogen absorption, stability and catalytic activity, however, leaching and coke formation have been substantial for catalysts in comparison to noble metal catalysts (Ooi, 2020).

1.7.2.4 Catalyst support

Support systems are another critical parameter for bio-oil hydroprocessing. Alumina, carbon, graphene, graphene oxides, TiO_2, ZrO_2 and CeO_2 are commonly used carriers (Dutta, Banerjee, & Jaiswal, 2020; Jaiswal, Sudhakar, & Ramaswamy, 2018). Most widely, alumina is used in the petroleum industry to remove sulphur. It has been seen that alumina is susceptible to attack by acidic water at elevated conditions, and subsequently, its reaction with water leads to a reduction in surface area. Alternatively, activated carbon and commercial carbon support are superior to traditional alumina support. The support with a high content of mesoporous carbon facilitated the diffusion of the reagents to the internal surface where the active sites were located. Silica is also an alternative due to its inert nature. MgO and ZrO_2 supports are also recognised as suitable essential supports for hydroprocessing reactions (Cordero-Lanzac, Rodríguez-Mirasol, Cordero, & Bilbao, 2021).

1.7.2.5 Temperature

The HDO of bio-oils is greatly affected by applied temperature. As the temperature rises over time, the degree of deoxygenation alters and considerably decreases the oxygen concentration in the bio-oils. At high temperatures, coke formation on the catalyst surface could be a primary concern and require further attention (Qu et al., 2021).

1.8 Catalytic synthesis of renewable diesel

Catalysts are essential elements that effectively impact the reaction process without significantly altering their composition or structure. Catalysts are mainly classified into three different categories: homogeneous, heterogeneous and enzymatic catalysts. Based on suitability and compatibility, catalysts are used in various reactions such as metathesis, carbonylation, oxidation, hydrogenation, hydrocyanation, etc. Homogeneous catalysts have been preferred, while heterogeneous catalysts have predominant uses in industrial applications (FT process, Haber–Bosch, etc.). Due to their easy recovery and reuse, heterogeneous catalysts are often considered for usage over homogeneous ones. Enzyme catalysts derived from plants, microbes and animal tissues have many applications (leather industry, dairy industry, paper and textile industry, detergent manufacturing, etc.) due to their easy natural availability and nontoxic nature (Lin, Amesho, Chen, Cheng, & Chou, 2020).

The choice of catalysts in the production of renewable biodiesel impacts numerous factors, including production yield and reaction rate. Sulphur, nonsulphur and transition metal catalysts are the catalysts commonly used to prepare renewable diesel. For example, nonsulphurised and sulphurised nickel catalysts are widely explored with different petroleum feedstocks to produce renewable diesel. Nickel or nickel-based materials (with other supports) are often used as heterogeneous catalysts and have good deoxygenation performance. It has been mentioned by Alvarez-Galvan, Campos-Martin, and Fierro (2019) that hydrogen gets trapped in nickel metal surface defects, and then the activation energy of hydrogen dissociation decreases. In addition, the electronic characteristics of the Ni metal can cause reactions analogous to those of noble metals such as platinum or palladium, such as the breaking of C—C or C—H bonds for the reaction of hydrocarbons (Alvarez-Galvan et al., 2019).

Nickel having an oxidation number of zero is the most functional form in terms of surface adsorption, activation and radical generating properties. Nickel catalysts and additional support materials such as activated carbon, zeolite and zirconia have been successfully used in renewable diesel generation. The accumulation of these support materials helps to improve acid–base sites; pore volume, particle size and surface area have a noteworthy influence on conversion rates with the composition of the prepared diesel. However, in the case of metal carbides, the catalytic selectivity is difficult to control because the thermal reaction generates various unwanted products and leads to the formation of different active phases of the catalyst during the hydrotreating process. Mainly, the carbide-related deoxygenation activity is decreased with increasing temperature due to a competitive hydrocracking reaction. Still, transition metal phosphides, molybdenum carbide, carbon-supported molybdenum carbide, molybdenum/tungsten carbides and Ru-supported alumina are some examples that have been investigated and mentioned in reports. Modifying factors such as carburising temperature and metal loading will help to improve catalytic activity in diesel production. Due to the high specific surface area and mesoporous structure, nickel/alumina and molybdenum or tungsten have also been suggested as good choices in recent studies. These catalysts have shown high catalytic activity and better yield of hydrocarbons using feedstocks such as UCO and sunflower oil (Derawi, Razak, & Asikin-Mijan, 2022).

1.9 Nanomaterial-based catalysts to produce renewable diesels

Catalysts play a vital role in the transesterification reaction of biodiesel synthesis. With the development of nanotechnology, researchers have used nanocatalysts with high catalytic performance to optimise both yield and energy consumption in biodiesel synthesis. The effectiveness of the catalytic activity in the process influences by certain parameters such as the percentage of load, the reaction time, the alcohol-to-oil ratio and the temperature. Compared with other nanocatalysts, CaO nanocatalysts have been widely used due to their abundance in nature. Recently, hybrid nanocatalysts derived from the organic–metallic framework have been frequently used in research, resulting in a wide-ranging catalytic activity to produce biodiesel. Nanomaterial-based catalysts for the production of renewable biodiesel have been illustrated in Fig. 1.2. Biodiesel is a fundamental component of our future energy supply in transportation and a promising substitute to fossil-based fuels with enormous environmental and socio-economic benefits. However, further advancement is the key to opening up biodiesel

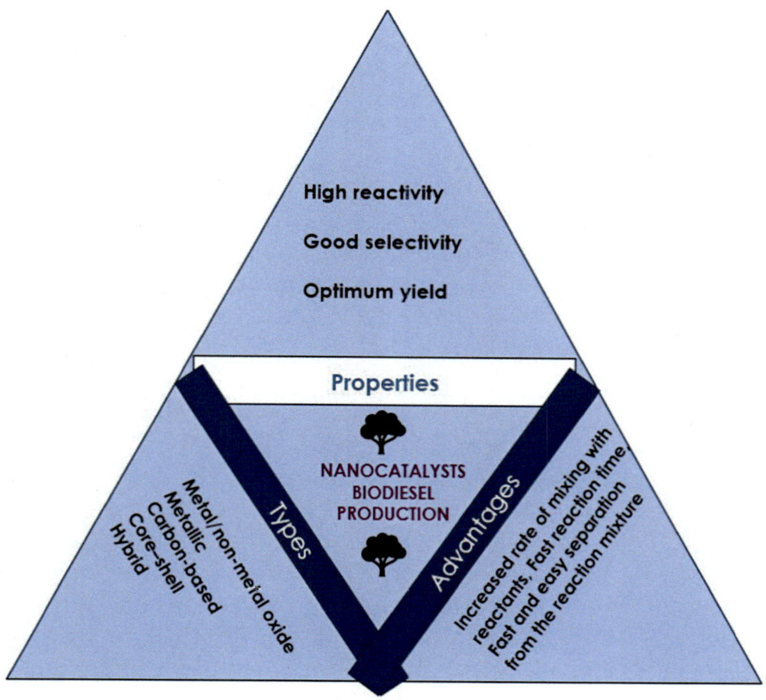

Figure 1.2 Nanomaterial-based catalysts to produce renewable biodiesel.

production more effectively and profitably. Nanotechnology plays a tremendous role in achieving these due to the desirable properties of nanocatalysts, such as higher surface area, durability, crystallinity, catalytic performance and energy storage capacity. Therefore nanocatalysts have been successfully used to optimise both energy and yield in biodiesel production. For instance, the high lignin content in plant-based materials limits the fermentable sugars extraction due to the inhibitor formations and the reduction of the metabolic processes (Zhang, Zhang, Zhou, Wu, & Xu, 2016).

Consequently, transesterification process becomes energy-intensive, and this problem has been addressed using nanocatalysts with a superior interfacial surface area, greater porosity, higher selectivity and effective active sites. Furthermore, except for enzymes, nanomaterials have greater chemical stability, better absorptive capacity, crystallinity and enhanced catalytic activities (Yang, Deng, Pan, Fu, & Bao, 2015). In general, the rate of chemical kinetics or catalytic activities and electron transfer are improved with nanomaterials due to the availability of a higher number of active sites due to the increased surface area/volume ratio. As a result, it reduces inhibitor formation and enhances the activities of anaerobic microbes to optimise biodiesel production. Not only that but also, the nanocatalysts immobilise the enzyme lipase for the production of biodiesel (Cherian, Kalavathy, Joshi, Phoebe, & Gurunathan, 2022).

Furthermore, the optimisation of biodiesel production using nanocatalysts has been governed by their size, morphology and geometry. The nanocatalysts used to produce biodiesel can be classified into acidic, basic and bifunctional nanocatalysts (Pandit & Mohapatra, 2021). So far, various organic, metallic or nonmetallic oxides (TiO_2, Fe_3O_4, ZnO, CaO), metallic (Ni, Ag, Au, Rh, Pd, Pt), carbon-based (graphene, fullerene, CNT, MWCNT), core–shell structures (Ag/SiO_2, Au/SiO_2, Ni/SiO_2, Fe_3O_4/SiO_2, Au/TiO_2, Fe/C, $FeNi/SiO_2$, ZnO/SiO_2) and hybrids (FeMo, CuMo) have been widely used in research. Moreover, the synthesis of nanocatalyst natural or waste raw materials has been widely used (Ahmad et al., 2022; Anr, Saleh, Islam, Hamdan, & Maleque, 2016; Kumar, Jaiswal et al., 2021; Kumar, Yogeshwar et al., 2021; Yang et al., 2022; Zulfiqar et al., 2021).

In recent research, seashells (*Polymedosa erosa*), cockle shells (*Anadara granosa*), *Boerhavia procumbens* leaves were used in the formation of CaO nanocatalysts (Abd Malek et al., 2021; Kumar, Jaiswal, et al., 2021; Kumar, Yogeshwar, et al., 2021). In particular, to obtain a high yield of biodiesel, researchers have used nanocatalysts of different sizes and concentrations. In addition, variations in reaction times, temperature and alcohol-to-oil ratios were applied to obtain an optimal condition, resulting in a high yield. In addition, the synthesis has been carried out by making variations in experimental parameters such as calcination temperature and load percentages. Although researchers have used pure nanocatalysts, hybrid nanocatalysts derived from an organic framework with a wide range of catalytic activity are extensively used in biodiesel production to maximise yield. Based on the results, it has been observed that experimental parameters (e.g. reaction time, temperature and catalyst loading) directly influence biodiesel yield. Furthermore, the catalytic activity of the nanocatalysts increased when the size became smaller forming more reactive sites due to the superior surface-to-volume ratio.

1.10 Physicochemical properties of diesel, biodiesel and renewable diesels

Determining the quality of diesel is essential to ensure good performance in applications. Numerous factors can impacts the quality of biodiesel, such as the quality of the feedstock (composition of fatty acids), production process and post-production parameter (handling and storage). The quality of diesel can be determined by analysing the physicochemical properties, such as viscosity, density, water content, cloud point (CP), pour point (PP) and heating value. Biodiesel produced by the transesterification process has many benefits over diesel. However, since most diesel engines have been designed to run on diesel, the physicochemical characteristics of biodiesel must be adapted to the designed engines (Barabás & Todoruţ, 2011). In addition, they are less polluting due to lower CO and particulate matter emissions. Furthermore, biodiesel is less toxic, biodegradable and has a lower risk of explosion than diesel due to its higher flashpoint. It is known that the oxygen required for the combustion of biodiesel is lower due to the availability of two oxygen atoms in its molecule. Improved lubrication in the ignition system can be considered another advantage that ultimately reduces wear and increases engine life (Vertonha, Delgado, & Lukasievicz, 2018). Here, the chapter deliberates the significant physicochemical characteristics of diesel, biodiesel and renewable diesel.

1.10.1 Cetane number

The CN can be defined as an index number representing the quality of diesel fuel and a measure of speed. In other words, the CN can be expressed as the percentage of hexadecane by volume in an ignitable composition of 1-methylnaphthalene and hexadecane, which resembles the characteristics of the diesel fuel to be tested. The acceptable range of CNs for diesel engine operation has been standardised between 40 and 55. A high CN represents easy and quick ignition of the fuel. It also counts the speed of self-ignition by diesel injection in a diesel engine (Bezergianni & Dimitriadis, 2013). In the case of biodiesel, the CN is not fixed. The carbon atoms responsible for forming the carbon chain, ester and unsaturated bonds are responsible for producing various factors in deciding the CN in biodiesel. This is mainly due to the mixture of multiple vegetable oils used to produce biodiesel. Biodiesel has been made from various plant-based oil from the feedstocks, such as karanja (*Millettia pinnata*), palm (*Elaeis oleifera*), peanut (*Arachis hypogaea*), jatropha (*Jatropha curcas*), rapeseed (*Brassica napus*), sunflower (*Helianthus annuus*) and soybean (*Glycine max*). The CN of vegetable oil ranges between 68 and 94.8, while in the case of animal fat, it ranges between 56 and 60. Renewable diesel can be defined as *n*-paraffinic hydrocarbons that do not include oxygen, sulphur and CN number ranging from 80 to 99. High CNs in the renewable diesel allow high-speed diesel engines to run efficiently. The CN in renewable diesel also depends on the long chain of hydrocarbons and the number of isomerised and short chains of hydrocarbons. The longer chain hydrocarbons dictate better CNs compared to isomerised ones (Dash, Chavan, Kumar, Ahamed, & Lingfa, 2020).

1.10.2 Density

Density can be defined as the quantity that resembles the amount of available matter that an object has in relation to its volume. The density of any substance can be defined as the ratio of the mass of a fluid to its volume. The density of any substance depends on temperature. The density of diesel is measured at a temperature ranging from 15°C to 40°C. The density of a substance correlates with fuel factors, CN, viscosity, arene structure, volatile nature and boiling point. In most cases, when the temperature of a substance increases, the density decreases while the volume increases. The density of biodiesel developed from palm oil, jatropha oil, rapeseed oil, mahua oil, etc., ranges between 800 and 890 kg/m^3. The density of renewable diesel ranges between 775 and 816 kg/m^3. According to EN 590, the density of renewable diesel fuel is lower than that of biodiesel (Al-Mohaimeed, Elshikh, Praveenkumar, & Sekar, 2021). This increases the pour rate by 4%–5% for diesel and biodiesel. According to Bernoulli's equation, the flow rate is inversely proportional to the square root of the fluid density. Similarly, renewable diesel and biodiesel fuel injected into a CI engine in the same quantity will have less volume in renewable diesel case due to the lower density. This also means that the renewable diesel will have lower energy content, showing less power and torque in the engine.

1.10.3 Cloud point

The CP of fuel represents the working condition in regions of lower temperatures. The cloud point can be defined as the temperature that allows the development of crystals by lowering fuel temperature. The formation of crystals thickens the fuel and gives it a cloudy appearance. This interferes with fuel filtration and fuel injectors. This also interferes with the operating capabilities of the fuel and hampers the work process in the diesel engine. The CP of vegetable oils used for biodiesel generation can be further reduced by hydroprocessing or transesterification processes. Biodiesel can efficiently work even at −5°C. The renewable diesel in these cases ranges from −30°C to −5°C. The CP of renewable diesel can be further lowered to -55°C through isomerisation. However, it is also observed that this same process reduces the CN of biodiesel. Short hydrocarbon chains allow better biodiesel flow properties than long hydrocarbon chains. Renewable diesel contains saturated bonds that would enable a better operation capacity. In the case of biodiesel, the CP is reduced due to the presence of saturated fatty esters (Bezergianni & Dimitriadis, 2013). Other properties such as PP and cold flow pour point are responsible for determining the functional properties of the fuel at lower temperatures.

1.10.4 Kinematic viscosity

Viscosity can be defined as the measure of friction between adjacent layers of fluid matter, such as gas and liquid, and that they resist movement. This resistance occurs between two adjacent layers of fluid. It is seen that the viscosity of biodiesel ranges from 4.0 to 4.8 mm^2/s, and that of renewable diesel falls between 3 and 4 mm^2/s at 40°C. Maintain proper fuel viscosity according to engine injector design. The appropriate blend of biodiesel provides better lubrication. The higher viscosity of biodiesel concerning standard values allows for poor atomisation and insufficient flow. This leads to oversized vapours during injection and inadequate and poor mixing resulting in incomplete combustion (Barabás & Todoruţ, 2011). Exhaust emissions such as CO, HC and PM increase with biodiesel consumption due to its low volatility and high viscosity compared to conventional diesel fuel. In the case of renewable diesel, emissions are reduced under similar working conditions. Lower viscosity leads to malformations in the fuel pumps. This increases the chances of the pump leaking, leading to insufficient fuel supply and engine destruction.

1.10.5 Oxygen content/stability

The molecular arrangement and the presence of oxygen play a vital role in determining the stability of biodiesel. Biodiesel comprises FAME which contains oxygen atoms and is aromatic. Renewable diesel is a gasoline-diesel that has linear saturated hydrocarbon chains without oxygen atoms and is aliphatic. Oxygen atoms are present in the molecular structure of the fuel, making it unstable and reactive. This makes biodiesel stability a challenge. The weight content of oxygen in renewable diesel is zero, while biodiesel is 6%–12%. This is an additional benefit to renewable diesel as it is long-lasting and stable compared to biodiesel (Azreena et al., 2022). Vegetable oils contain a fair amount of double-bonded fatty acids that interfere with oxygen stability. The stability of the oxygen also depends on the storage time of the biodiesel. Oxygen stability interferes with several properties in biodiesel. The addition of antioxidants such as *tert*-butylated hydroxytoluene, pyrogallol, *tert*-butylated hydroxylquinone, propyl gallate and *tert*-butylated hydroxyanisole has been shown to improve oxidation stability. Studies have shown that among these pyrogallol is more efficient. Biodiesel is recommended away from exposure to light, atmospheric oxygen and water vapours to ensure more extended stability. Heteroatom removal is initiated by hydroprocessing. Hydroprocessing can be classified as HDS, HDO and hydrodenitrogenation with the heteroatom removal efficiency as HDS > HDO > HDN (Leckel, 2006).

1.10.6 Flashpoint

Flashpoint can be defined as the minimum temperature at which an organic compound gives off sufficient vapour for ignition under exposure to flame or fire. Flashpoint represents the volatility and flammability of biodiesel. It is essential to take safety precautions. The vapour pressure of different fluids is different, which determines the temperature of the fluid and its responsibility with respect to Boyle's law. The increase in temperature also raises its vapour pressure and expands the vapour concentration (Fu, 2019). This also indicates that temperature plays a crucial role in determining the formation of vapour concentration. The amount of vapours emitted at the flashpoint is not always sufficient to support combustion. It would only initiate ignition whenever the fuel is in direct contact with a fire source. The fuel volatility is inversely proportional to its flashpoint; the lower the flashpoint, the higher its volatility.

1.10.7 Calorific value

Calorific value can be defined as the amount of energy that a substance can contain or release during combustion. The calorific value in biodiesel ranges from 33.6 to 43.2 MJ/kg, while that of renewable diesel ranges from 42 to 47 MJ/kg. The existence of O_2 in the atomic structures of the oil reduces the calorific value of biodiesel by 10%–12%. Hydroprocessing initiates the removal of oxygen atoms, which helps improve its calorific value (Koul, Kumar, & Singh, 2021).

1.11 Industrial application of renewable diesel

Neste Oil Corporation in Singapore has developed NExBTL technology to produce more than 800,000 tons of renewable diesel per year from feedstocks. The renewable diesel produced has a high cetane value between 84 and 99, a low CP value (as low as −30°C) and increased stability over a more extended period. These properties make renewable diesel the best option for improving engine capacity (Hasan & Rahman, 2017). The UOP Ecofining Honeywell Green Diesel Unit in Alabama, United States, has developed a hydrotreating process to produce renewable diesel from animal fats, greases and oils (Glisic, Pajnik, & Orlović, 2016). The Tyson Foods Inc. plant in Geismar, LA, United States, has developed a new technology to produce green diesel by hydrotreating nonedible grade animal fats, such as chicken fat, beef tallow and pork lard. Tyson and Syntroleum Corporation have collaborated to develop a technology to produce synthetic fuels targeted at the green diesel, military fuel and jet fuel markets. They are trying to manufacture up to 75 million gallons of renewable diesel fuel per year.

Haldor Topsoe developed a new hydrotreating technology to produce green diesel and jet fuel from raw inedible tallows. The

Preem Gothenburg Sweden refinery has adopted this technology. ConocoPhillips meets the need for renewable diesel to European Union standards at the Whitegate Refinery in Cork, Ireland, by producing 365,000 barrels per year of environmentally friendly diesel fuel for sale in the Irish market. They are producing sulphur-free renewable diesel through the hydrogenation process from vegetable oil. This renewable diesel is used to blend and transport petro-diesel. At the St Charles Refinery in Norco, LA, Valero Energy Corporation's plant developed hydrogenation and isomerisation of UCO and animal fat to produce renewable diesel. The joint venture Toyota Motor Corporation (TMC), Hino Motors, the Tokyo Metropolitan Government and Nippon Oil Corporation (NOC) began in 2005 to develop a second-generation renewable diesel fuel, bio-hydro-refined diesel. They produce it through the hydrogenation process using vegetable oil feedstock (Norouzi & Talebi, 2020; Vasquez, Silva, & Castillo, 2017). Various investigations into renewable diesel production from nonedible feedstocks have developed solutions to environmental crises and global fuel demand. Renewable diesels are dignified to perform a significant role in the modern economy as production costs get lower. In the future, together with the production of renewable diesel, its commercialisation also becomes important. Countries around the globe need to promote the production of renewable fuels at an industrial level.

1.12 Prospective future and recommendations

Future prospects for renewable diesel synthesis depend on new technologies and cost-effectiveness. It also depends on new research exploring the potential of renewable diesel to maintain and improve the sustainability of energy use. The generation of biofuels from the hydrolysis of lignocellulosic biomass is promising for the future of energy. It is ecological and renewable. The breakthrough in the commercial production of renewable diesel coupled with subsiding enzyme prices has also contributed positively to the advancement of renewable diesel synthesis. More research is needed to overcome existing challenges in the field. Some of the prospects involved in this include exploration of new renewable diesel resources, process improvement, process affordability, preparation of a suitable cocktail for biomass and final formulation of the conversion process. The use of enzymes in the 'saccharification' of biomass in the pretreatment process can be a prospective research area for improving the synthesis of renewable hydrocarbon biofuels.

1.13 Conclusion

Biofuel has a lot of potential to change the face of the energy crisis and expenses in the world. It can be utilized for lighting, heating, transportation and other energy-related processes. It has the potential to rejuvenate the agricultural and energy sectors. It can be concluded here that renewable diesel, especially biodiesel, is a complex and advancing area. Alternative fuels are gaining in popularity and importance over time. Alternative fuels have successfully replaced petroleum-based fuels and contributed to sustainability. The power of carbon renewability is improved compared to petroleum-based fuel. Biofuel production must meet the requirements of the 'Renewable Fuels Standard (RFS)' under the 'Energy Independence and Security Act of 2007'. RFS also monitors the percentage of lignocellulosic ethanol produced from biomass. Biodiesel can be commercially manufactured via HDO and transesterification processes of vegetable oils, fats and other waste cooking oils. Ultrasonication, mechanical agitation and supercritical alcohol are used to generate FAME biodiesel. Several processes have also been used to develop renewable diesel, such as the white diesel process, the green diesel process, the hybrid diesel process and the FT diesel process. Hydrotreating, thermal conversion and liquid biomass are the three most essential steps in renewable diesel production. Catalysts also play a vital role in the conversion process and yield increase. It also contributes to saving energy for the high-temperature conversion process. Liquid catalysts are also used for the generation of FAME biodiesel. However, there are some concerns regarding the availability of a large amount of feedstock for biofuel production.

References

Abd Malek, M. N. F., Pushparaja, L., Hussin, N. M., Embong, N. H., Bhuyar, P., Rahim, M. H. A., & Maniam, G. P. (2021). Exploration of efficiency of nano calcium oxide (CaO) as catalyst for enhancement of biodiesel production. *Journal of Microbiology, Biotechnology and Food Sciences, 11*(1), e3935, e3935.

Abdulkareem-Alsultan, G., Asikin-Mijan, N., Lee, H. V., Rashid, U., Islam, A., & Taufiq-Yap, Y. H. (2019). A review on thermal conversion of plant oil (edible and inedible) into green fuel using carbon-based nanocatalyst. *Catalysts, 9*(4), 350.

Ahmad, M., Elnaggar, A. Y., Teong, L. K., Sultana, S., Zafar, M., Munir, M., ... Abidin, S. Z. U. (2022). Sustainable and eco-friendly synthesis of biodiesel from novel and non-edible seed oil of *Monotheca buxifolia* using green nanocatalyst of calcium oxide. *Energy Conversion and Management: X, 13*, 100142.

Alalwan, H. A., Alminshid, A. H., & Aljaafari, H. A. (2019). Promising evolution of biofuel generations. Subject review. *Renewable Energy Focus, 28*, 127–139.

Ali, E. N., & Tay, C. I. (2013). Characterization of biodiesel produced from palm oil via base catalyzed transesterification. *Procedia Engineering, 53*, 7–12.

Al-Mohaimeed, A. M., Elshikh, M. S., Praveenkumar, T. R., & Sekar, M. (2021). Exergy and energy analysis of α-Fe_2O_3 doped Al_2O_3 nanocatalyst based biodiesel blends—Performance and emissions characteristics. *Journal of Energy Resources Technology*, 1–32.

Alvarez-Galvan, M. C., Campos-Martin, J. M., & Fierro, J. L. (2019). Transition metal phosphides for the catalytic hydrodeoxygenation of waste oils into green diesel. *Catalysts, 9*(3), 293.

Ambaye, T. G., Vaccari, M., Bonilla-Petriciolet, A., Prasad, S., van Hullebusch, E. D., & Rtimi, S. (2021). Emerging technologies for biofuel production: a critical review on recent progress, challenges and perspectives. *Journal of Environmental Management, 290*, 112627.

Amin, A. (2019). Review of diesel production from renewable resources: Catalysis, process kinetics and technologies. *Ain Shams Engineering Journal, 10*(4), 821–839.

Anr, R., Saleh, A. A., Islam, M. S., Hamdan, S., & Maleque, M. A. (2016). Biodiesel production from crude Jatropha oil using a highly active heterogeneous nanocatalyst by optimizing transesterification reaction parameters. *Energy & Fuels, 30*(1), 334–343.

Arora, N., Jaiswal, K. K., Kumar, V., Vlaskin, M. S., Nanda, M., Pruthi, V., & Chauhan, P. K. (2020). Small-scale phyco-mitigation of raw urban wastewater integrated with biodiesel production and its utilization for aquaculture. *Bioresource Technology, 297*, 122489.

Asikin-Mijan, N., Lee, H. V., Abdulkareem-Alsultan, G., Afandi, A., & Taufiq-Yap, Y. H. (2017). Production of green diesel via cleaner catalytic deoxygenation of Jatropha curcas oil. *Journal of Cleaner Production, 167*, 1048–1059.

Azreena, I. N., Lau, H. L. N., Asikin-Mijan, N., Hassan, M. A., Izham, S. M., Gamal, M. S., . . . Taufiq-Yap, Y. H. (2022). Hydrodeoxygenation of fatty acid over La-modified HZSM5 for premium quality renewable diesel production. *Journal of Analytical and Applied Pyrolysis, 161*, 105406.

Barabás, I., & Todoruţ, I. A. (2011). Biodiesel quality, standards and properties. *Biodiesel-Quality, Emissions and By-products*, 3–28.

Baskar, G., & Soumiya, S. (2016). Production of biodiesel from castor oil using iron (II) doped zinc oxide nanocatalyst. *Renewable Energy, 98*, 101–107.

Bezergianni, S., & Dimitriadis, A. (2013). Comparison between different types of renewable diesel. *Renewable and Sustainable Energy Reviews, 21*, 110–116.

Bhatia, S. K., Bhatia, R. K., Jeon, J. M., Pugazhendhi, A., Awasthi, M. K., Kumar, D., . . . Yang, Y. H. (2021). An overview on advancements in biobased transesterification methods for biodiesel production: Oil resources, extraction, biocatalysts, and process intensification technologies. *Fuel, 285*, 119117.

Bjelić, A., Grilc, M., & Likozar, B. (2020). Bifunctional metallic-acidic mechanisms of hydrodeoxygenation of eugenol as lignin model compound over supported Cu, Ni, Pd, Pt, Rh and Ru catalyst materials. *Chemical Engineering Journal, 394*, 124914.

Cherian, E., Kalavathy, G., Joshi, T.J., Phoebe, M.G. L., & Gurunathan, B. (2022). *Importance of nanocatalyst and its role in biofuel production*. In Gurunathan, Baskar, Sahadevan, Renganathan, & Zakaria, Zainul Akmar. (Eds.), Biofuels and bioenergy (pp. 171–182). Elsevier.

Choudhary, M., Joshi, S., Rao, L., & Srivastava, N. (2021). *Production of biofuel from disposed food and dairy waste.* In Thatoi, Hrudayanath, Mohapatra, Sonali, & Das, Swagat Kumar. (Eds.), Bioprospecting of enzymes in industry, healthcare and sustainable environment (pp. 123–138). Singapore: Springer.

Chrysikou, L. P., Dagonikou, V., Dimitriadis, A., & Bezergianni, S. (2019). Waste cooking oils exploitation targeting EU 2020 diesel fuel production: Environmental and economic benefits. *Journal of Cleaner Production, 219,* 566–575.

Cordero-Lanzac, T., Rodríguez-Mirasol, J., Cordero, T., & Bilbao, J. (2021). Advances and challenges in the valorization of bio-oil: Hydrodeoxygenation using carbon-supported catalysts. *Energy & Fuels, 35*(21), 17008–17031.

Dash, S. K., Chavan, S. B., Kumar, A., Ahamed, M. S., & Lingfa, P. (2020). *Jatropha biodiesel blends as renewable diesel fuel additives. Bioresource utilization and bioprocess* (pp. 93–105). Singapore: Springer.

D'Cruz, A., Kulkarni, M. G., Meher, L. C., & Dalai, A. K. (2007). Synthesis of biodiesel from canola oil using heterogeneous base catalyst. *Journal of the American Oil Chemists' Society, 84*(10), 937–943.

de Onís, C. M. (2021). *Energy islands* (1st, pp. 1–300). University of California Press, ISBN: 9780520380639.

Derawi, D., Razak, N.A. A., & Asikin-Mijan, N. (2022). *Advanced biofuels: Deoxygenation and hydrodeoxygenation catalytic reaction.* In Nanda, Sonil, & Vo, Dai-Viet. (Eds.), Innovations in thermochemical technologies for biofuel processing (pp. 219–237). Elsevier.

Douvartzides, S. L., Charisiou, N. D., Papageridis, K. N., & Goula, M. A. (2019). Green diesel: Biomass feedstocks, production technologies, catalytic research, fuel properties and performance in compression ignition internal combustion engines. *Energies, 12*(5), 809.

Dutta, S., Banerjee, I., & Jaiswal, K. K. (2020). Graphene and graphene-based nanomaterials for biological and environmental applications for sustainability. *Octa Journal of Biosciences, 8*(2), 106–112.

Dutta, S., Jaiswal, K. K., Verma, R., Basavaraju, D. M., & Ramaswamy, A. P. (2019). Green synthesis of zinc oxide catalyst under microwave irradiation using banana (*Musa* spp.) corm (rhizome) extract for biodiesel synthesis from fish waste lipid. *Biocatalysis and Agricultural Biotechnology, 22,* 101390.

Folayan, A. J., Anawe, P. A. L., Aladejare, A. E., & Ayeni, A. O. (2019). Experimental investigation of the effect of fatty acids configuration, chain length, branching and degree of unsaturation on biodiesel fuel properties obtained from lauric oils, high-oleic and high-linoleic vegetable oil biomass. *Energy Reports, 5,* 793–806.

Fu, J. (2019). Flash points measurements and prediction of biofuels and biofuel blends with aromatic fluids. *Fuel, 241,* 892–900.

Glisic, S. B., Pajnik, J. M., & Orlović, A. M. (2016). Process and techno-economic analysis of green diesel production from waste vegetable oil and the comparison with ester type biodiesel production. *Applied Energy, 170,* 176–185.

Hancsók, J., Krár, M., Magyar, S., Boda, L., Holló, A., & Kalló, D. (2007). Investigation of the production of high cetane number bio gas oil from pre-hydrogenated vegetable oils over Pt/HZSM-22/Al_2O_3. *Microporous and Mesoporous Materials, 101*(1–2), 148–152.

Hasan, M. M., & Rahman, M. M. (2017). Performance and emission characteristics of biodiesel–diesel blend and environmental and economic impacts of biodiesel production: A review. *Renewable and Sustainable Energy Reviews, 74,* 938–948.

Heidari, N., & Pearce, J. M. (2016). A review of greenhouse gas emission liabilities as the value of renewable energy for mitigating lawsuits for climate change related damages. *Renewable and Sustainable Energy Reviews, 55*, 899–908.

Huber, G. W., O'Connor, P., & Corma, A. (2007). Processing biomass in conventional oil refineries: Production of high quality diesel by hydrotreating vegetable oils in heavy vacuum oil mixtures. *Applied Catalysis A: General, 329*, 120–129.

Jaiswal, K. K., Banerjee, I., & Mayookha, V. P. (2021). Recent trends in the development and diversification of sericulture natural products for innovative and sustainable applications. *Bioresource Technology Reports, 13*, 100614.

Jaiswal, K.K., Dutta, S., Banerjee, I., Mayookha, V.P., & Bhushan, M. (2022). *Lipid extraction from fish processing residues for sustainable biofuel production.* In Galanakis, Charis. (Ed.), Sustainable fish production and processing (pp. 293–319). Academic Press.

Jaiswal, K. K., Dutta, S., Banerjee, I., Pohrmen, C. B., & Kumar, V. (2021). Photosynthetic microalgae–based carbon sequestration and generation of biomass in biorefinery approach for renewable biofuels for a cleaner environment. *Biomass Conversion and Biorefinery*, 1–19.

Jaiswal, K. K., Jha, B., & Prasath, R. A. (2014). Biodiesel production from discarded fish waste for sustainable clean energy development. *Journal of Chemical and Pharmaceutical Sciences, 974*, 2115.

Jaiswal, K. K., Kumar, V., Vlaskin, M. S., & Nanda, M. (2020). Impact of glyphosate herbicide stress on metabolic growth and lipid inducement in *Chlorella sorokiniana* UUIND6 for biodiesel production. *Algal Research, 51*, 102071.

Jaiswal, K. K., Kumar, V., Vlaskin, M. S., & Nanda, M. (2021). Impact of pyrene (polycyclic aromatic hydrocarbons) pollutant on metabolites and lipid induction in microalgae *Chlorella sorokiniana* (UUIND6) to produce renewable biodiesel. *Chemosphere, 285*, 131482.

Jaiswal, K. K., & Pandey, H. (2014). Next generation renewable and sustainable micro-fuels from *Chlorella pyrenoidosa. International Journal of Recent Scientific Research, 5*(4), 767–769.

Jaiswal, K. K., & Prasath, A. R. (2016). Integrated growth potential of *Chlorella pyrenoidosa* using hostel mess wastewater and its biochemical analysis. *International Journal of Environmental Sciences, 6*(5), 592–599.

Jaiswal, K. K., Sudhakar, S., & Ramaswamy, A. P. (2018). Graphene'–World's thinnest material for revolutionizing applications. *Everyman's Science, 53*, 219–223.

Kaewmeesri, R., Srifa, A., Itthibenchapong, V., & Faungnawakij, K. (2015). Deoxygenation of waste chicken fats to green diesel over Ni/Al_2O_3: Effect of water and free fatty acid content. *Energy & Fuels, 29*(2), 833–840.

Kim, J., Sen, S. M., & Maravelias, C. T. (2013). An optimization-based assessment framework for biomass-to-fuel conversion strategies. *Energy & Environmental Science, 6*(4), 1093–1104.

Kiss, A. A., & Bildea, C. S. (2012). A review of biodiesel production by integrated reactive separation technologies. *Journal of Chemical Technology & Biotechnology, 87*(7), 861–879.

Koul, R., Kumar, N., & Singh, R. C. (2021). A review on the production and physicochemical properties of renewable diesel and its comparison with biodiesel. *Energy Sources, Part A: Recovery, Utilization, and Environmental Effects, 43*(18), 2235–2255.

Kumar, V., Arora, N., Pandey, S., Jaiswal, K. K., Nanda, M., Vlaskin, M. S., & Chauhan, P. K. (2020). Microwave-assisted pretreatment of harmful algal blooms for microbial oil-centered biorefinery approach. *Biomass Conversion and Biorefinery*, 1–9.

Kumar, V., Jaiswal, K. K., Tomar, M. S., Rajput, V., Upadhyay, S., Nanda, M., ... Kurbatova, A. (2021). Production of high value-added biomolecules by microalgae cultivation in wastewater from anaerobic digestates of food waste: A review. *Biomass Conversion and Biorefinery*, 1–18.

Kumar, Y., Yogeshwar, P., Bajpai, S., Jaiswal, P., Yadav, S., Pathak, D. P., ... Tiwary, S. K. (2021). Nanomaterials: Stimulants for biofuels and renewables, yield and energy optimization. *Materials Advances*, *2*(16), 5318–5343.

Leckel, D. (2006). Catalytic hydroprocessing of coal-derived gasification residues to fuel blending stocks: Effect of reaction variables and catalyst on hydrodeoxygenation (HDO), hydrodenitrogenation (HDN), and hydrodesulfurization (HDS). *Energy & Fuels*, *20*(5), 1761–1766.

Leung, D. Y. C., & Guo, Y. (2006). Transesterification of neat and used frying oil: Optimization for biodiesel production. *Fuel Processing Technology*, *87*(10), 883–890.

Limayem, A., & Ricke, S. C. (2012). Lignocellulosic biomass for bioethanol production: Current perspectives, potential issues and future prospects. *Progress in Energy and Combustion Science*, *38*(4), 449–467.

Lin, Y. C., Amesho, K. T., Chen, C. E., Cheng, P. C., & Chou, F. C. (2020). A cleaner process for green biodiesel synthesis from waste cooking oil using recycled waste oyster shells as a sustainable base heterogeneous catalyst under the microwave heating system. *Sustainable Chemistry and Pharmacy*, *17*, 100310.

Long, F., Liu, W., Jiang, X., Zhai, Q., Cao, X., Jiang, J., & Xu, J. (2021). State-of-the-art technologies for biofuel production from triglycerides: A review. *Renewable and Sustainable Energy Reviews*, *148*, 111269.

Madden, S., Alles, K., & Demirel, Y. (2021). Measuring sustainability of renewable diesel production using a multi-criteria decision matrix. *Biofuels, Bioproducts and Biorefining*, *15*(6), 1621–1637.

Mutreja, V., Singh, S., & Ali, A. (2011). Biodiesel from mutton fat using KOH impregnated MgO as heterogeneous catalysts. *Renewable Energy*, *36*(8), 2253–2258.

Niu, M., Ji, P., Fan, Z., Yan, Y., Li, D., & Li, W. (2020). Hydrofining process of coal tar based on four kinds of catalyst grading. *Energy & Fuels*, *34*(5), 6510–6517.

Norouzi, N., & Talebi, S. (2020). An overview on the green petroleum production. *Chemical Review and Letters*, *3*(1), 38–52.

Ooi, X.Y. (2020). *Production of renewable hydrocarbon fuel via titanium oxide-alumina (TiO_2-Al_2O_3) catalytic deoxygenation reaction/Ooi Xian Yih* [Doctoral dissertation]. Universiti Malaya.

Pandit, P.R., & Mohapatra, A. (2021). *Calcium-based nanocatalysts in biodiesel production.* In Ingle, Avinash P. (Ed.), Nano- and biocatalysts for biodiesel production, (pp. 93–113), Wiley.

Pasupulety, N., Gunda, K., Liu, Y., Rempel, G. L., & Ng, F. T. (2013). Production of biodiesel from soybean oil on CaO/Al_2O_3 solid base catalysts. *Applied Catalysis A: General*, *452*, 189–202.

Patel, M., & Kumar, A. (2016). Production of renewable diesel through the hydroprocessing of lignocellulosic biomass-derived bio-oil: A review. *Renewable and Sustainable Energy Reviews*, *58*, 1293–1307.

Pattanaik, B. P., & Misra, R. D. (2017). Effect of reaction pathway and operating parameters on the deoxygenation of vegetable oils to produce diesel range hydrocarbon fuels: A review. *Renewable and Sustainable Energy Reviews*, *73*, 545–557.

Pelemo, J., Inambao, F., & Onuh, E. (2020). Potential of used cooking oil as feedstock for hydroprocessing into hydrogenation derived renewable diesel: A review. *International Journal of Engineering Research and Technology*, *13*, 500–519.

Peng, B., Yuan, X., Zhao, C., & Lercher, J. A. (2012). Stabilizing catalytic pathways via redundancy: Selective reduction of microalgae oil to alkanes. *Journal of the American Chemical Society, 134*(22), 9400–9405.

Qu, L., Jiang, X., Zhang, Z., Zhang, X. G., Song, G., Wang, H., ... Chang, Y. L. (2021). A review of hydrodeoxygenation of bio-oil: Model compounds, catalysts, and equipment. *Green Chemistry, 23*(12). Available from https://doi.org/10.1039/D1GC03183J.

Ren, X. Y., Feng, X. B., Cao, J. P., Tang, W., Wang, Z. H., Yang, Z., ... Zhao, X. Y. (2020). Catalytic conversion of coal and biomass volatiles: A review. *Energy & Fuels, 34*(9), 10307–10363.

Shaah, M. A. H., Hossain, M. S., Allafi, F. A. S., Alsaedi, A., Ismail, N., Ab Kadir, M. O., & Ahmad, M. I. (2021). A review on non-edible oil as a potential feedstock for biodiesel: Physicochemical properties and production technologies. *RSC Advances, 11*(40), 25018–25037.

Shamoon, A., Haleem, A., Bahl, S., Javaid, M., Garg, S. B., Sharma, R. C., & Garg, J. (2022). Environmental impact of energy production and extraction of materials—A review. *Materials Today: Proceedings.*

Shirasaki, Y., Nasu, H., Hashimoto, T., & Ishihara, A. (2019). Effects of types of zeolite and oxide and preparation methods on dehydrocyclization-cracking of soybean oil using hierarchical zeolite-oxide composite-supported Pt/NiMo sulfided catalysts. *Fuel Processing Technology, 194*, 106109.

Singh, D., Subramanian, K. A., & Garg, M. O. (2018). Comprehensive review of combustion, performance and emissions characteristics of a compression ignition engine fueled with hydroprocessed renewable diesel. *Renewable and Sustainable Energy Reviews, 81*, 2947–2954.

Smagala, T. G., Christensen, E., Christison, K. M., Mohler, R. E., Gjersing, E., & McCormick, R. L. (2013). Hydrocarbon renewable and synthetic diesel fuel blendstocks: Composition and properties. *Energy & Fuels, 27*(1), 237–246.

Speight, J. G. (2004). New approaches to hydroprocessing. *Catalysis Today, 98*(1–2), 55–60.

Tabatabaei, M., Aghbashlo, M., Dehhaghi, M., Panahi, H. K. S., Mollahosseini, A., Hosseini, M., & Soufiyan, M. M. (2019). Reactor technologies for biodiesel production and processing: A review. *Progress in Energy and Combustion Science, 74*, 239–303.

Taromi, A. A., & Kaliaguine, S. (2018). Green diesel production via continuous hydrotreatment of triglycerides over mesostructured γ-alumina supported NiMo/CoMo catalysts. *Fuel Processing Technology, 171*, 20–30.

Tuli, D., & Kasture, S. (2022). *Biodiesel and green diesel.* In Tuli, Deepak, Kasture, Sangita, & Kuila, Arindam. (Eds.), Advanced biofuel technologies (pp. 119–133). Elsevier.

Upare, D. P., & Upare, P. P. (2020). Biomass-derived triglyceride: A source of renewable aviation fuel and biodiesel. *Alternative Energy Resources*, 193–222.

Vasquez, M. C., Silva, E. E., & Castillo, E. F. (2017). Hydrotreatment of vegetable oils: A review of the technologies and its developments for jet biofuel production. *Biomass and bioenergy, 105*, 197–206.

Vertonha, M. T., Delgado, L. A., & Lukasievicz, G. V. B. (2018). Characterization of the physicochemical properties of different biodiesel samples. *Brazilian Archives of Biology and Technology*, 61.

Wang, Y., Ou, S., Liu, P., Xue, F., & Tang, S. (2006). Comparison of two different processes to synthesize biodiesel by waste cooking oil. *Journal of Molecular Catalysis A: Chemical, 252*(1–2), 107–112.

Wen, L., Wang, Y., Lu, D., Hu, S., & Han, H. (2010). Preparation of KF/CaO nanocatalyst and its application in biodiesel production from Chinese tallow seed oil. *Fuel, 89*(9), 2267–2271.

Wildschut, J., Mahfud, F. H., Venderbosch, R. H., & Heeres, H. J. (2009). Hydrotreatment of fast pyrolysis oil using heterogeneous noble-metal catalysts. *Industrial & Engineering Chemistry Research, 48*(23), 10324–10334.

Yang, C. M., Huynh, M. V., Liang, T. Y., Le, T. K., Huynh, T. K. X., Lu, S. Y., & Tsai, D. H. (2022). Metal-organic framework-derived Mg-Zn hybrid nanocatalyst for biodiesel production. *Advanced Powder Technology, 33*(1), 103365.

Yang, F., Deng, D., Pan, X., Fu, Q., & Bao, X. (2015). Understanding nano effects in catalysis. *National Science Review, 2*(2), 183–201.

Zhang, L., Zhang, L., Zhou, T., Wu, Y., & Xu, F. (2016). The dual effects of lignin content on enzymatic hydrolysis using film composed of cellulose and lignin as a structure model. *Bioresource Technology, 200*, 761–769.

Zhou, L., & Lawal, A. (2015). Evaluation of presulfided NiMo/γ-Al_2O_3 for hydrodeoxygenation of microalgae oil to produce green diesel. *Energy & Fuels, 29*(1), 262–272.

Zhou, L., & Lawal, A. (2016). Hydrodeoxygenation of microalgae oil to green diesel over Pt, Rh and presulfided NiMo catalysts. *Catalysis Science & Technology, 6*(5), 1442–1454.

Zikri, A., & Aznury, M. (2020). *Green diesel production from crude palm oil (CPO) using catalytic hydrogenation method, . IOP conference series: Materials science and engineering* (Vol. 823, p. 012026). IOP Publishing, April.

Zulfiqar, A., Mumtaz, M. W., Mukhtar, H., Najeeb, J., Irfan, A., Akram, S., … Nabi, G. (2021). Lipase-PDA-TiO2 NPs: An emphatic nano-biocatalyst for optimized biodiesel production from *Jatropha curcas* oil. *Renewable Energy, 169*, 1026–1037.

2

Standard specifications for renewable diesel

Umer Rashid[1,2], Chuan Li Lee[3], Balkis Hazmi[1], Safa Gamal[4,5] and Mahnoush Beygisangchin[1]

[1]Institute of Nanoscience and Nanotechnology, Universiti Putra Malaysia, Serdang, Selangor, Malaysia [2]Center of Excellence in Catalysis for Bioenergy and Renewable Chemical (CBRC), Faculty of Science, Chulalongkorn University, Bangkok, Thailand [3]Institute of Tropical Forestry and Forest Product, Universiti Putra Malaysia, Serdang, Selangor, Malaysia [4]Putra Laboratory for Sustainable Bioenergy and Chemicals, Faculty of Science, Universiti Putra Malaysia, Serdang, Selangor, Malaysia [5]Department of Chemistry, Faculty of Science, Universiti Putra Malaysia, Serdang, Selangor, Malaysia

2.1 Introduction

The diminishing reserves of fossil fuels and environmental pollution attributed to their consumption have spurred research on generating fuels and energy from other sources, such as wind, solar and biomass (Gholizadeh, Hu, & Liu, 2019). Biomass mainly refers to materials primarily composed of agricultural and forestry wastes. It is a carbonaceous substance that can be used as a renewable feedstock for obtaining syngas, liquid fuels and solid tar through thermochemical processes such as pyrolysis. Biomass can be transformed into a condensed liquid known as bio-oil by pyrolysis at temperatures from 350°C to 650°C in an inert environment (Xu, Li, Tu, Sun, & Jiang, 2018). Bio-oil is a blend of organic compounds, some of which are with added value. Thus, rather than producing biofuels from bio-oil, the manufacture of chemicals such as polyols, carboxylic acids, phenolic acids and levulinates has attracted interest (Gholizadeh et al., 2019). Alternatively, biomass wastes can also be converted into biofuel, which can be divided into four generations based on the biomass feedstock they use: edible biomass, non-edible biomass, algal biomass and chemical processing. The conversion of biomass into fuels, chemicals and materials is extremely important because it allows waste to be recycled while also increasing the added value of the entire industrial chain (Li et al., 2021).

Renewable Diesel. DOI: https://doi.org/10.1016/B978-0-323-91153-5.00004-2

The plantation of first-generation biofuel sources only recorded at 5% of the world's arable land, which cannot accommodate the demands for food security and fuel production, is often blamed for an increase in vegetable oil prices (Singh, Guldhe, Rawat, & Bux, 2014). Biofuels made from food and plantation waste, as well as non-food crops, can avoid the food *vs* fuel debate (Tian, Tai, Chua, & Li, 2017). The triglyceride feedstocks used in biodiesel synthesis are catalytically hydroprocessed to produce renewable diesel fuel, also known as green diesel (Mikkonen, 2008). These alternative fuels are probably best referred to as "renewable diesel." Renewable fuels, as an alternative fuel, can be produced from several resources via a hydrodeoxygenation reaction at high temperatures and pressures in the presence of a catalyst (Amin, 2019; Knothe, 2010). However, all of the alternative diesel-like fuels derived from biological resources can be considered as "renewable," albeit to a varying degree. In this chapter, renewable diesel-like fuels such as biodiesel, green diesel (hydroprocessed vegetable oil), Fischer–Tropsch (FT) diesel, white diesel and hybrid diesel are discussed. The search for renewable diesel has resulted in the development of fuels derived from a variety of sources, including renewable feedstocks such as fats and oils.

There is no need for alcohol in the hydroprocessing process, the products are hydrocarbons rather than fatty acid alkyl esters, and there is no glycerol residue. Biodiesel and renewable diesel are referred to collectively as "biodistillate." However, the problems caused by the high oxygen content in biodiesel could be alleviated mainly via hydrodeoxygenation. Nevertheless, up to now, the process is still costly due to the requirements for externally added hydrogen, the use of the metal-oxide catalyst for hydrogenation and their rapid deactivation induced by coking (Zacher, Olarte, Santosa, Elliott, & Jones, 2014).

2.2 Renewable diesel feedstocks

Renewable diesel seems to be a better alternative for addressing the issue of rapidly depleting fossil fuels and the limitations of biodiesel. In the presence of excess hydrogen, the hydroprocessing of liquid biomass at an elevated temperature of approximately 300°C–400°C and high pressure of approximately 5–20 MPa over a catalyst such as nickel-molybdenum, cobalt-molybdenum, palladium-carbon, platinum-carbon, ruthenium, etc. supported on alumina and zeolite materials yields a drop-in fuel known as renewable diesel (Koul, Kumar, & Singh, 2021).

Renewable diesel is produced with the same goal in mind as biodiesel production: to cut down on CO_2 emissions while also providing cleaner combustion than petroleum diesel. Crude oil reserves are not found in all countries. In fact, the cost of importing oil has a significant impact on importers of oil. A country's reliance on fossil fuels can be reduced if more people start switching to biofuels. With the growth of the biofuel industry, more jobs will be created, ensuring the stability of our economy. Currently, Malaysia's biofuel policies have demonstrated the Malaysian government's efforts to reduce reliance on fossil fuels (Chia et al., 2022). To bridge the gaps from the conventional use of fossil feedstock to improve various conversion processes, it is crucial to gradually integrate renewable solutions into mass production. This is notably applicable to recent developments in the use of renewable feedstock and energy resources in reforming processes. Despite all of the benefits associated with biofuels, they are currently quite expensive to produce. Currently, interest and capital investment in biofuel production are low. The cost of research (including intellectual property rights), stringent emission norms and future installation will result in a significant increase in the price of biofuel. Researchers are working hard to improve methods for extracting biofuel at a lower cost. As of now, the prices are comparable to gasoline and are still feasible (mostly without taxes). The discussion on these topics provides researchers with information and critical discussion on renewable diesel production, as well as keeps the related communities up to date on the latest developments from both industrial and academic research.

The advantages of renewable diesel, such as its high heating value, similar energy density to petroleum diesel, exceptional storage stability and non-corrosiveness, make it a better alternative for an enhanced source of energy (Linganiso et al., 2022). Vegetable oil, animal fats, waste cooking oil (WCO) and lignocellulosic biomass are some of the feedstock used to produce renewable diesel, which is similar to biodiesel (Tuli & Kasture, 2022). Among these, vegetable oil was discovered to be a compelling alternative that is not only renewable but also simple to produce. These oils derived from various plants and seeds play an important role in the production of biodiesel to replace diesel, despite the fact that they have their own set of issues when used directly as fuel in a diesel engine (Bukkarapu, 2019). The demand for suitable biofuel crops has risen because of growing biofuel mandates, giving the agriculture sector a boost. If all of these feedstocks are readily accessible on a global scale, an excellent strategic framework can be developed to organise the cycle of biofuel collection, purchasing, biodiesel production and supply to end users while prioritising economic energy and a clean environment (Yaqoob et al., 2021). Regrettably, the processing of biodiesel is much more expensive than that of fossil diesel. The price of biodiesel ranges between 1.5 and 2 times that of diesel fuel (Gholami, Pourfayaz, & Maleki, 2021). Thus, much research has been done to find an economically viable production process using various feedstock in order to reduce the cost of biodiesel. Given the similarity of the feedstock used to produce biodiesel and renewable diesel, different types of feedstocks are being researched and studied in order to produce high-quality renewable diesel, as shown in Tables 2.1 and 2.2.

2.3 Diesel-like alternative fuels

Numerous commercial plants for producing diesel-like alternative fuels have been built around the world. The most straightforward approach is to use oil derived from vegetable and animal fats to produce renewable diesel. It includes, but is not limited to, used restaurant oil, soyabean oil, jatropha oil, algae oil and other similar substances. These oils are primarily composed of triglycerides and free fatty acids (Farooqui, Kumar, Sinha, & Ray, 2022). Hydrotreating lipids and biodiesel can yield renewable diesel. Renewable diesel is a cleaner fuel than biodiesel and has better cold flow properties (Tyagi, Aslam, & Sarma, 2022). Moreover, diesel-like alternative fuel with superior properties to biodiesel, has attracted considerable attention and interest in commercialisation (Munyentwali, Li, & Yang, 2022). As reported by Kelly (2021), the United States produced 533 million gallons of diesel-like alternative fuel in 2020, with a capacity of 2.65 billion gallons predicted by 2024. Diesel-like alternative fuel has also been produced through thermal decomposition. However, technologies for converting feedstock into diesel-like alternative fuel are still in the early stages of development. Moreover, the products generated by all routes must be upgraded to meet international fuel standards. The sections that follow go into greater detail about the other type renewable diesel; for instance, biodiesel (fatty acid methyl ester, FAME), green diesel, FT diesel, white diesel and hybrid diesel.

2.4 Biodiesel

Biodiesel is a part of alternative and renewable biofuels for replacing fossil-based diesel. Before the presence of biodiesel processing technologies, peanut oil was directly tested in the

Table 2.1 Characteristics of renewable diesel feedstocks.

Feedstock	Positive impact	Negative impact	References
Jatropha oil	Hydroprocessed fuel can reduce greenhouse gas emissions by up to 75% when compared to Jet-A fuel. High oil yield (55%–60% by mass). Better cold flow properties than diesel. Calorific value is greater than that of biodiesel. Viscosity and density are comparable to diesel. Fruit shell and seed husk can be valorised for biogas production. The production cost is low as it can grow on relatively low water and nutrients.	Long chain of hydrocarbon with a lower percentage of oxygen. Their ability to be used as biodiesel is hampered by high viscosity factor. High engine wear and NOx emissions than diesel fuel. Poor oxidation stability. Longer gestation period (3–5 years) than other oilseed crops.	Alherbawi, McKay, Mackey, and Al-Ansari (2021), Ewunie, Morken, Lekang, and Yigezu (2021), Koul, Kumar, and Singh (2022), Liu et al. (2015), Singh et al. (2021), Yadav, Singh, Singh, and Negi (2021)
Canola oil	Highly desirable with low density, iodine index, acid number (0.071) and viscosity. Relatively cheaper and easily obtained. It is safer to transport and store as its flash point is higher than diesel. High oxygen content capable of shortening the ignition delay time, improving the combustion environment and causing the fuel to burn more completely, reducing carbon monoxide, particulate matter and other exhaust emissions.	Not preferred for use in engines unless further modified.	Bharti et al. (2021), Chinasamy (2021), Ge, Yoon, and Choi (2017), Gholami et al. (2021)
Palm oil	Carbon dioxide, carbon monoxide and tetrahydrocannabinol emissions are significantly lower when compared to regular diesel combustion.	High temperature and long reaction time required in the hydroprocessing process. Higher extraction costs in comparison to fossil fuels.	Bangjang, Kaewchada, and Jaree (2021), Faizal and Ateeb (2018), Mekhilef, Siga, and Saidur (2011), Ruangudomsakul et al. (2021)

(*Continued*)

Table 2.1 (Continued)

Feedstock	Positive impact	Negative impact	References
	Higher yield, low fertiliser, water and pesticide requirement for the plantation. Safe to handle, any spill will be easier and less expensive to clean up, which is not the case with fossil fuels.	High production cost, fluctuating crude palm oil prices result in low or negative profit margins.	
Sunflower oil	Great adaptability to wider spatial variability. Easily accessible and non-toxic in nature. Relatively inexpensive when compared to other edible oils. The seeds yield 48%—52% oil, with a high proportion of polyunsaturated fatty acids.	Low volatility High viscosity has a negative effect on atomisation quality, resulting in poor engine performance and exhaust emission which may causes engine parts to fail. Pumping oil from the tank and injecting from the injectors is difficult, and large drops form, making combustion difficult. Higher pour point of its methyl ester precludes its use in unmodified diesel engines operating in cold temperatures.	Attaphaiboon, Neramittagapong, Neramittagapong, and Theerakulpisut (2021), Bukkarapu (2019), Fayad, Chaichan, and Dhahad (2021), Ilkiliç and Yücesu (2005), Tahir (1982), Tutunea et al. (2019)
Soybean oil	Higher engine output by up to 3%, with lower values of hydrocarbons, carbon monoxide, smoke and particulate matter in exhaust. The combustion parameters of soybean oil methyl ester, namely ignition delay, peak pressure and rate of pressure rise, were similar to those of diesel fuel.	High emission of NOx.	Bharti et al. (2021), Gangolu, Chunchu, Yallamati, and Gopidesi (2021), Gopinath, Puhan, and Nagarajan (2010)
Mahua oil	The seed contains approximately 35%—40% oil, whereas the kernel contains approximately 70% oil. Each tree produces between 20 and 40 kg of seed. Fruit can be produced by an adult tree for up to 60 years.	NOx emissions are higher in comparison to diesel. The amount of phospholipids in solvent-extracted oil is twice that of pressed oil.	Kumar and Raheman (2021), Kumar and Simhadri (2021), Rathore and Ramchandani (2016)

(Continued)

Table 2.1 (Continued)

Feedstock	Positive impact	Negative impact	References
Corn oil	Mostly contains unsaturated FAMEs (about 85%). The presence of mostly unsaturated fatty acid methyl esters improves cold flow properties such as viscosity, pour point and cloud point. Less expensive than other vegetable oils.	Resulting in poor thermal oxidation. Low inherent oxidative stability, with a polyunsaturated content of more than 50%. Emit more nitrogen oxides and have lower thermal efficiency. More density, higher viscosity/	Doğan and Naktiyok (2022), Muruganantham, Pandiyan, and Sathyamurthy (2021), Rama Krishna Reddy et al. (2022)
Waste cooking oil	Carbon monoxide, hydrocarbon and other emissions were lower than in diesel fuel. Performance is comparable to diesel fuel. Smoke opacity is about 60% less than petroleum diesel. It does not require any additional lubricants (unlike diesel), due to its excellent lubricating properties.	High viscosity and incompatible chemical property of the acid number require modification. Corrosion and increased engine wear caused by high acidity value. Higher exhaust gas temperatures for engine load. Heating value is lower than diesel. Low calorific value, reducing the thermal efficiency of the powered engine. Temperature rise does not favour conversion (90% at 330°C).	Abed, El Morsi, Sayed, Shaib, and Gad (2018), Bezergianni, Dimitriadis, Kalogianni, and Pilavachi (2010), Chinasamy (2021), Yaqoob et al. (2021)

Table 2.2 Fuel properties of renewable diesel feedstock and biodiesel derived from renewable feedstock.

Type of feedstock	Density (kg/L)	Cetane number	Acid value (mg KOH/g)	Heating value (MJ/kg) approx	Cloud point (°C)	Flash point (°C)	Pour point (°C)	Kinematic viscosity (mm^2/s)	References
Jatropha oil	0.91	37.83	28.00	39.1	6.3	142	3	4.1	Dugala, Goindi, and Sharma (2021), Ibrahim, Abed, Gad, and Abu Hashish (2020), Ibrahim, Asikin-Mijan, et al. (2020), Koul et al. (2021)
Jatropha biodiesel	0.87	42.62	0.49	38.8	2.3	121	−3.0	1.4	Ibrahim, Abed et al. (2020), Ibrahim, Asikin-Mijan, et al. (2020), Koul et al. (2021), Dugala et al. (2021)
Canola oil	0.88	55.00	0.25	n/a	−9.7	172	−21.7	4.5	Alviso, Zárate, and Duriez (2021), Binhweel et al. (2021)
Canola biodiesel	0.86	54.30	0.48	38.6	−3.3	178	−8.0	4.4	Binhweel et al. (2021)
Palm oil	0.91	45.50	10.04	37.4	22.2	219	7.0	2.8	Ekop, Etuk, and Eddy (2007), Mehta and Anand (2009), Pioch and Vaitilingom (2005)
Palm biodiesel	0.85	61.00	0.96	38.3	18.0	183	6.7	6.8	Goga et al. (2018), Mehta and Anand (2009), Peter et al. (2020)
Sunflower oil	0.92	38.80	0.52	39.6	2.5	274	−15.5	33.9	Alviso et al. (2021), Goga et al. (2018), Saydut et al. (2010)
Sunflower biodiesel	0.88	50.00	0.12	38.5	−14.0	178	−16.0	4.3	Goga et al. (2018), Mehta and Anand (2009)
Soybean oil	0.92–0.93	39.50	1.01	37.6	−6.0	270	−9.8	31.5	Eke-Ejiofor, Beleya, and Allen (2021), Ma'arof et al. (2021), Mehta and Anand (2009), Stanciu (2018), Subroto (2020)
Soybean oil biodiesel	0.87–0.89	57.60	0.04–0.37	45.1	−1.0	171–182	−6.0	4.1–6.0	Alviso et al. (2021), Khanal and Shah (2021), Neuana, de Sousa Barboza, dos Santos, and da Silva (2021)

(Continued)

Table 2.2 (Continued)

Type of feedstock	Density (kg/L)	Cetane number	Acid value (mg KOH/g)	Heating value (MJ/kg) approx	Cloud point (°C)	Flash point (°C)	Pour point (°C)	Kinematic viscosity (mm^2/s)	References
Mahua oil	0.92	52.30	34.00	35.0–36.0	12.0	235	8.0	24.7	Alviso et al. (2021), Dugala et al. (2021), Guharaja et al. (2016), Padhi and Singh (2010), Puhan, Nagarajan, Vedaraman, and Ramabramhmam (2007)
Mahua biodiesel	0.870	52.00	0.54	39.91	7.6	195	2.7	4.2	Dugala et al. (2021), Puhan et al. (2007)
Corn oil	0.919	40.20	0.65	37.56	−1.9	n/a	−22.4	30.8	Alajtal et al. (2018), Alviso et al. (2021), Liu, Driskell, and Engelhardt (2006), Stanciu (2020)
Corn biodiesel	0.840	54.80	0.11	37.46	−9.0	146	−13.0	4.6	Doğan and Naktiyok (2022), Muruganantham et al. (2021)
Waste cooking oil (canola oil)	0.87	61.00	2.10	42.95	−1.0	164–167	−16.0	5.0	Chhetri, Watts, and Islam (2008), Ebenaza and Vinoth (2015), Ibrahim, Asikin-Mijan, et al. (2020)
Waste cooking oil biodiesel	0.865	65.19	0.56	37.71	−6.0	120	−10.0	1.5	Ibrahim, Abed, et al. (2020), Ibrahim, Asikin-Mijan, et al. (2020), Koul et al. (2021), Saydut et al. (2010)

diesel engine in late 19th century (Singh et al., 2020). However, it was found to be less effective and impractical due high viscosity, poor atomisation, engine deposits, acidic nature, susceptibility towards oxidation and resultant solidification etc. Thus, countless research has been done for converting the vegetable oil into biodiesel fuel to overcome the stated problems by four methods such as microemulsion, thermal cracking, blending and transesterification. Among those methods, transesterification is the most preferable biodiesel production method due to its simplicity and low cost which is basically the mixing of oil or fat with alcohol in the presence of catalyst. The utilisation of catalyst such as alkali catalyst, acidic catalyst, bifunctional acid–basic catalyst and enzymatic catalyst should be bsed on the feedstock properties. The catalysed reaction would convert the feedstock fatty acid into its derived esters such as myristate (C14:0), palmitate (C16:0), stearate (C18:0), oleate (C18:1) and linoleate (C18:2) which can be identified and measured by gas chromatogram (Ibrahim, Asikin-Mijan, et al., 2020). In addition, every feedstock contains different percentage of fatty acid as listed in Table 2.3.

2.4.1 Biodiesel standard specification

The fuel properties of synthesised biodiesel must comply with the prevailing specifications such as American Society for Testing and Materials (ASTM) D6571 (United States) and EN 14214 (European Union, EU) for better performances, control of noxious emissions and handling safety. Table 2.4 lists the qualities of synthesised biodiesel that were produced by WCO of different regions along with the standards/limits recommended by EN and ASTM standards. As reflected in Table 2.4, biodiesel with low viscosity would provide good injection flow and atomisation of fine droplets for the complete combustion (Hazmi, Rashid, Kawi, Mokhtar, et al., 2022). It is also important to measure the moisture content as water can enhance corrosion inside engine compartments, and can also reduce calorific value (Silitonga et al., 2013). The lowest temperature at which biodiesel will ignite when exposed to a source of ignition/spark is known as the flash point. The stated flash point were slightly higher than standard which indicated good handling, storage and transportation safety (Yadav & Sharma, 2019). Cloud point and pour point represent the lowest temperature for biodiesel to form cloud crystal and halt from flowing owing to solidification. Cetane number of prepared biodiesels were recorded within standard limit by the and indicating good compression combustion engine performance of low knocking.

Table 2.3 Fatty acid composition of biodiesel feedstocks.

Feedstock	FFA, %	$C_{14:0}$	$C_{16:0}$	$C_{18:0}$	$C_{18:1}$	$C_{18:2}$	Biodiesel yield, %	References
WCO–KOH	1.39	1.40	34.92	4.04	42.55	13.57	96.72	Farid et al. (2020)
Palm fatty acid distillate (PFAD) $SO_3H–ZnAl_2O_4$	90.0	1.93	46.68	7.9	40.19	4.25	94.6	Soltani, Rashid, Nehdi, Al-Resayes, and Al-Muhtaseb (2017)
Chicken fat $CaO/CuFe_2O_4$	–	11.8	25.58	11.86	36.52	6.21	94.5	Seffati, Honarvar, Esmaeili, and Esfandiari (2019)
Jatropha oil $CaSO_4/Fe_2O_3$-SiO_2	6.80	–	20.16	7.22	39.77	31.53	94.0	Teo et al. (2019)
Karanja oil	2.53	–	11.65	7.50	51.59	16.46	–	Sahoo and Das (2009)
Fat oil grease H_2SO_4	12	1.3	38.3	7.2	36.9	15.4	97.4	Montefrio, Xinwen, and Obbard (2010)
WCO $ZnAl_2O_4$	5.61	–	30.5	5.2	24.8	39.1	94.9	Shaban et al. (2020)

2.5 Green diesel

As stated by Tuli and Kasture (2022) green diesel is a second-generation biofuel with properties that are superior to those of petroleum-based diesel but with a similar molecular structure. Green diesel could also be manufactured from biomass via a thermochemical route using biomass sugars. The high similarity to fossil fuels confers several advantages to the use of green diesel. The main benefits of green diesel over biodiesel are the ability to use it in much higher proportions in mixtures with conventional diesel and the ability to produce it using the existing infrastructure used in the production of diesel in a petroleum refinery. Green diesel comprises a mixture of the saturated straight chain and branched hydrocarbons, which usually comprising around 15–18 carbon atoms (C_{15} to C_{18}). There is a resemblance of this composition with that of a fossil petroleum diesel, thus enabling the green diesel to be used as a pure form in compression ignition (CI) engines, or blended with any required ratio without the need to modify the engine (Douvartzides, Charisiou, Papageridis, & Goula, 2019). A hydroprocessing catalytic reaction, as well as/or either a deoxygenation reaction (deCOX) involving decarboxylation/decarbonylation ($DeCO_2$/DeCO) of biomass-based feedstocks, are used to produce the green diesel (Moreira, Rezende, & Pasa, 2020; Safa-Gamal et al., 2021).

Table 2.4 Physicochemical parameters and standards for biodiesel.

Properties	ASTM D6571	EN 14214	Synthesised biodiesel (WCO-based)			
			Hazmi, Rashid, Kawi, Nur, et al. (2022)	Shaban et al. (2020)	Abukhadra and Sayed (2018)	Rashid et al. (2013)
Viscosity, cSt	1.9–6.0	3.5–5.0	3.95	5.15	3.87	4.17
Moisture, wt.%	<0.05	<0.05	0.01	–	0.03	–
Flash point, °C	>120	>120	160	117	128	164
Cloud point, °C	−3 to 15	[a]	5	−12	6	10
Pour point, °C	−5 to 10	[a]	3	0	5	8
Cetane number	47–65	51–120	60	–	53	58
Density, g/cm^3	0.82–0.90	0.86–0.90	0.82	0.82	0.86	0.88

[a]Not specified. EN 14214 uses time and location-dependent values for the cold filter plugging point (CFPP) instead.

2.5.1 Green diesel fuel specifications

The green diesel specification prescribes the required properties of diesel fuel and sets the limits (requirements) for the values of these properties. These requirements are listed in Table 2.5 defined by the specification. The specification also contains the standard test methods used to measure the values of the properties. The specification contains the minimum mandatory requirements needed to guarantee acceptable performance for most users. In addition, this specification recognises some requirements established by the EPA to reduce emissions.

For the development of diesel fuel and diesel-powered cars to be successful, refiners, engine and vehicle manufacturers, as well as other interested parties, need to agree on the qualities of diesel fuel required for adequate performance and dependable operation. This consensus has been reached earlier as in the case of ASTM D6751 and EN 14214. Specifications for diesel fuels have also been developed by the EU. With adjustments to meet local supply, crudes and regulations, these specifications are widely utilised in the nations of the Pacific basin and Asia.

Green diesel is oxygen-free, thus resulting in a higher heating value as well as increased energy density. The quantity of heating energy released when a particular amount of a substance undergoes complete combustion in the presence of oxygen under typical conditions for temperature and pressure is known as the heat of combustion of a substance, also known as the calorific value or the energy value. For diesel fuel, a greater calorific value of roughly 45 MJ/kg is preferred. Light hydrocarbons

Table 2.5 Comparison of petroleum diesel and green diesel specification (Patel & Kumar, 2016).

Properties	Test methods	Petroleum diesel ASTM D-975-17a	Green diesel EN 15940:2018
Molecular structure	–	Mixed hydrocarbons	Saturated/unsaturated alkanes
Cetane index, min	ASTM D-976	40	70–90
Kinematic viscosity @ 40°C, cSt	ASTM D-445	–	–
Density, g/mL	ASTM D-4052	0.82–0.845	0.78–0.85
Sulphur, ppm	ASTM D-5453	>10	<10
Cloud point, °C	ASTM D-2500	−5	(−25) – 30
Pour point, °C	ASTM D-97	Baseline	Excellent
Oxidative stability	EN15751		
Flash point, °C	ASTM D-93	52–136	68–120

reduce the calorific value, whereas heavier hydrocarbons have no discernible impact on the quantity of heat of combustion (Doustdar, Wyszynski, Mahmoudi, & Tsolakis, 2016). Moreover, green diesel is characterised by a high cetane number (70–90) as a result of its high alkane content (Ameen, Azizan, Yusup, Ramli, & Yasir, 2017; Safa-Gamal et al., 2021). The cetane number refers to a characteristic of diesel fuel that is used in measuring the propensity of igniting sprayed fuel into a standard tested engine apparatus. Thus, increased cetane values are indicative that a more intense ignition of fuel will occur timely.

The poor characteristics of low temperature, as this occurs in a chilly climate where crystallisation starts with tiny, solid visible crystals in the fuel known as the cloud point promptly, is one of the severe problems associated with the use of green diesel (Dwivedi & Sharma, 2013; Singh et al., 2020). The cloud point of green diesel is greater than petroleum diesel's due to heavy alkane components; however, this attribute can be rectified by mixing readily with any amount of petroleum diesel to use in various conditions (Zhao, Wei, Cheng, & Julson, 2017). In addition, flash point relate to the temperature wherein the formation of fuel can lead to a combustible mixture when exposed to flame, it is determined for safety in handling and use. The flash point of the green diesel is similar to petroleum diesel; therefore, it is more compatible with fuel as compared to biodiesel.

Diesel fuel long-term storage stability has been estimated using oxidation stability experiments. For this purpose, a number of oxidation stability test methods are currently in use, including Rancimat (EN15751) and Petro OXY (EN16091). It is common for these tests to have an elevated temperature and then added oxygen or air to accelerate the oxidation of the test fuel, and hence accelerate conduction of the test. Additionally, it has been suggested that these oxidation stability tests could be used to estimate the thermal stability of the diesel fuels. Stability of EN590 B0 (winter and summer grade) as well as the effect of cetane improver was evaluated with different oxidation stability methods (Hartikka, Kiiski, Kuronen, & Mikkonen, 2013). Generally, green diesel might be a more preferred renewable product due to the presence of an identical structure in molecules as compared to conventional petroleum diesel (Kittisupakorn, 2016).

2.6 Fischer–Tropsch diesel

The FT process produces chemicals, gasoline and diesel fuel. Since most FT products are linear, the quality of the diesel fuel is excellent, with cetane numbers up to 75. Generally, coal, methane and biomass are the three principal starting materials used in FT process. In nations with an excess of natural gas, methane serves as the primary raw material (Malaysia, Qatar, Nigeria, etc.). A new method created by Nippon GtL Technology Research Association uses the CO_2 present in natural gas to perform FT process without using oxygen. The technique is currently being developed at the commercial scale through additional demonstration tests. To convert syngas into hydrocarbons, the catalyst must be able to activate the carbon–oxygen link and remove oxygen atoms. In practical applications, the metals Ni, Co and Fe can be used to disrupt the C–O bond. Cobalt is typically applied to produce waxes and diesel fuel, while iron is used for gasoline production. For waxes, low temperatures (180°C–250°C) are utilised, while high temperatures (300°C–340°C) are used for the creation of gasoline (dos Santos & Alencar, 2020; Rauch, Kiennemann, & Sauciuc, 2013). The FT process can be used to produce synthetic fuels from bioenergy sources. The fact that FT-derived fuels are drop-in replacements for fossil fuels with fewer greenhouse gas emissions, low levels of sulphur and nitrogen, and low particulate emissions are some of the highlights of the technique.

Practically, PerkinElmer Clarus 60 Gas Chromatography–Mass Spectrometry (GC–MS) was used to monitor and analyse the hydrocarbon distribution of FT process in the liquid phase of products. According to GC analysis, the components of diesel (C_{12}–C_{22}) make up 75% of the liquid products, of which 97% are paraffin. Olefins and alcohols make up the remaining components (Doustdar et al., 2016). A decrease in olefin production and an increase in the percentage of heavier hydrocarbons in the product results in increased lubricity, as shown by the hydrocarbon distribution of F–T process on liquid products.

2.6.1 Fischer–Tropsch product specifications

A blend of roughly C_{10}–C_{22} straight-chain hydrocarbons that satisfies the cetane number, flash point, distillation profile and cloud point requirements set down by local regulators is referred to as "green diesel". However, FT fuels have a low density by nature, making them unable to meet diesel standards as shown in Table 2.6. Thus on-spec distillate is a better name for the gasoline created by this procedure (i.e. meeting all specifications of diesel excluding density). The product attributes were computed to confirm that these requirements were met using vales of flash point, cetane number, cloud point and distillation

Table 2.6 Specification of green diesel produced in United States and European Union (Doustdar et al., 2016).

Properties	Green diesel ASTM D-975-17b (United States)	Green diesel EN 940:2014 (EU)
Cetane Index, min	49	51
Kinematic viscosity @ 40°C, cSt	1.9–4.1	2.0–4.5
Density, g/mL	–	0.820
Sulphur, ppm	>10	<10
Cloud point, °C	−5	−10
Water, ppm	500	200
Distillation, T_{95}	370	360
Flash point, °C	55	55

curve. Moreover, FT LTFT distillate generally has extremely little sulphur and aromatics. Aspen PlusPensky-Martens's flash point estimation (ASTM D-93) was used to determine flash points (Tucker, Bordoloi, & van Steen, 2021).

2.7 White diesel

White diesel is the product of catalytic hydrotreatment of 100% WCO and consists of a single catalytic hydrotreatment step. White diesel technology was developed and demonstrated in the pilot-plant facilities of the Chemical Process and Energy Recourses Institute (CPERI) of the Research and Technology Hellas (CERTH). The white diesel technology was demonstrated via the BIOFUELS-2G environmental project as it was employed for the operation of a regular garbage truck from the fleet of the Municipality of Thessaloniki, Greece (Bezergianni & Dimitriadis, 2013).

2.7.1 White diesel fuel specifications

White diesel is a paraffinic fuel, free of sulphur and aromatics. As catalytic hydrotreatment targets the heteroatom removal, white diesel has negligible sulphur levels of ~1.54 ppm (Table 2.7). Moreover, it has a low density of 0.79 g/mL, which is lower compared to diesel and biodiesel standards (Teresa L. Alleman, 2016). This means that a higher fuel volume must be supplied for the same amount of energy compared to diesel. However, as a lower

Table 2.7 Properties of white diesel (Bezergianni, Dimitriadis, & Voutetakis, 2011).

Property	Units	Value
Density @ 60°C	g/mL	0.7532
Sulphur content	mg/kg	1.54
Nitrogen content	mg/kg	1.37
Cetane index		77
Flash point	°C	116
Water content	mg/kg	0
Viscosity @ 40°C	cSt	3.54
Total acid number	mg KOH/g	0
Copper strip corrosion (3 h in 50°C)		1b
Induction time/Oxidation stability @ 110°C	h	>22
Distillation (90%)	°C	303
Net heating value	MJ/kg	49
Cold filter plugging point (CFPP)	°C	20
Pour point	°C	23

density fuel, it has the potential to reduce the level of smoke. The cetane number of a white diesel is very high and other properties are very similar to other biofuels. White diesel has increased oxidation stability of >22 hours (Bezergianni & Chrysikou, 2012), compared to diesel and biodiesel standards, which means that white diesel does not need antioxidant additives to meet the satisfactory level of oxidation stability like diesel fuels with low levels of natural antioxidants. Furthermore, white diesel is also very safe to use and has significantly high flash point (116°C). The cold flow properties of white diesel are its major drawback, which can be overcome either by a second isomerisation step (which will increase its production cost) or by blends with conventional diesel.

The fuel properties of white diesel are given in the following Table 2.7. This new renewable diesel shows remarkable properties. It is a light diesel with no impurities (S, N, O, H_2O). It has an impressive cetane index which exceeds 77 (compared to fossil diesel with values ~50) and a net heating value of 49 MJ/kg (compared to conventional fossil diesel with 42–43 MJ/kg). Furthermore, even though white diesel is a bio-based fuel, it shows high oxidation stability (>22 hours while the EN standard is 6 hours) and also negligible acidity. As white diesel is highly paraffinic it exhibits problems with its cold flow properties which needs to be addressed in downstream processing through additives.

2.8 Hybrid diesel

Hybrid fuels are basically referred to as vegetable oils and some other added preservatives. Vegetable oil is much more viscous than either petroleum diesel or biodiesel. Hybrid fuels are a renewable source of energy, but their direct injection depends upon the viscosity and volatility of the fuel. The vegetable oils, however, cannot be directly used because of their properties. The high viscosity that is present in the fuel interferes in the operational process and hence leads to the poor results. To overcome these problems various techniques are employed such as dilution and blending of the oils with diesel. In addition, transesterification and microemulsifiers are used to make hybrid fuel. Formation of microemulsions ensures successful mixing of the vegetable oil and ethanol by lowering the surface tension between the two phases so that one stable homogenous mixture can be obtained. The surfactant actually adjusts itself in such a way that the hydrophilic end is oriented towards the water and the hydrophobic towards the oil (Kumar, 2014). It is highly required that such hybrid diesel fuel compositions be thermodynamically stable microemulsions, and are able to undergo a number of manipulations like pumping, storage and other similar activities with minimal risk of splitting. The technical problem at stake was then that of providing hybrid diesel fuel compositions which were less water-incompatible, bearing in mind, also, the circumstance that water is a diesel-fuel improver, in that it lowers the combustion temperature and reduces both the emission of fumes and the formation of nitrogen oxides.

Patent US-A-4 451 267 has suggested using a surface-active agent in order to improve diesel fuel compositions, while EP-A-0 117 915 has suggested adopting higher alcohols to stabilise blends of conventional diesel fuel. In order to offset the drawbacks of the prior art approaches outlined above, a hybrid diesel fuel composition contains a diesel fuel and fuel oil. Fuel oil, as is well known, is a by-product of the alcoholic fermentation of starches and sugars and is obtained by the redistillation of the crude ethanol. A reportedly typical fuel oil composition comprises, on a weight basis, 38% of isoamyl alcohol, 25% of isobutyl alcohol, 4.5% of isopropyl alcohol 13% of ethanol, 0.5% of methanol and other components, plus 19% of water. Refined fuel oil could be used, of course, but it is not recommendable from an economic standpoint.

Besides hydroprocessing of vegetable oil, in order to produce diesel-range hydrocarbons, there is an option of vegetable oil co-processing with petroleum-derived raw materials. This solution allows utilisation of existing refinery technologies, that is catalytic hydroprocessing and equipment by integrating liquid biomass into typical refining operations for the production of hybrid fuels. Several types of vegetable oil can be employed for co-processing with petroleum streams such as soybean, rapeseed, coconut, palm, jatropha oil and WCO (Bezergianni & Dimitriadis, 2013).

2.8.1 Hybrid diesel standard specifications

Afew co-processing studies show that the vegetable oil component increases the overall hybrid diesel yield. Hybrid diesel density ranges from 0.78 to 0.85 g/mL while the 90% (v/v) distillate temperature is within 300°C–330°C. Moreover, kinematic viscosity varies within diesel specifications (2.7–5.5 cSt), cetane number is significantly high (50–101) as is also the net heating value (43.3–47 MJ/kg). The cold flow properties are also quite diverse based on the catalyst and operating parameters employed. Generally, pour point is between −20°C and 26°C while cloud point ranges between −23°C and 20°C. Furthermore, the hybrid diesel is also low sulphur (3–13 ppm) and aromatics free (0.1–1.2 wt.%) fuel, thus they can be considered "clean fuels." Finally, catalytic co-hydrocracking technology may be more suitable as the cracking, treating and isomerisation reactions result in a hybrid diesel fuel of better cold flow properties (Bezergianni & Dimitriadis, 2013).

2.9 Comparison of renewable diesel

The world's population is expected to reach 12 billion by 2030 which will also increase the annual energy needs of the developing and resource-constrained countries. Energy demand in developing countries is expected to increase by 90%, with renewable energy sources such as biofuels accounting for approximately one-third of this increase (Vignesh, Pradeep Kumar, Ganesh, Jayaseelan, & Sudhakar, 2021). Despite the constant development of new low-carbon fuels, two major groups of renewable fuels in the diesel pool have emerged: biodiesel and green diesel. Green diesel uses the same feedstock as biodiesel, but it is processed through hydrotreating, thermal conversion, or biomass-to-liquid processes (Chia et al., 2022; Morrissey, Thoma, & López, 2021). Green diesel which is chemically identical to diesel can be used as a drop-in replacement in all current diesel engines. It is a diesel substitute of renewable origin (vegetable oils and fats). Green diesel should not to be confused with biodiesel as it is produced via catalytic hydroprocessing of vegetable oils and fats and not transesterification (Fernández-Villamil & Paniagua, 2018). Unlike biodiesel, green diesel does not contain oxygen, making it more stable, non-corrosive and comparable to petroleum diesel in terms of heating value. Green diesel also outperforms biodiesel in cold weather, emits fewer nitrogen oxides and has a higher cetane number, making CI engines easier to start. Finally, propane, a gaseous fuel with a high market value, is a by-product of green diesel produced by hydro-processing of triglycerides. It makes green diesel production more cost-effective in comparison to the production of biodiesel (Douvartzides et al., 2019; Kordouli, Sygellou, Kordulis, Bourikas, & Lycourghiotis, 2017; Kordouli, Kordulis et al., 2017).

Biodiesel, also known as FAME, is an alternative fuel derived from vegetable oils that can be used in place of petroleum diesel fuel (Ahadan, Adipoetra, Nuryati, Dewi, & Nasikin, 2021). Biodiesel is typically produced by transesterifying vegetable oils or animal fats with methanol or ethanol to produce mono-alkyl esters of long-chain fatty acids (Hachemane, 2022; Vlnieska, Muniz, Oliveira, César-Oliveira, & Kunka, 2022). As biodiesel contains oxygen atoms, it differs chemically from regular petroleum diesel. Most diesel engine manufacturers, like ethanol blending limits in gasoline engines, only support diesel blends up to B20, with higher blends requiring fuel system modifications (Afzal et al., 2021; Agarwal, 2007). Since biodiesel is sulphur-free, there are fewer SOx emissions and a reduction in particulate matter in the exhaust. Due to the absence of sulphur in biodiesel, it aids in reducing the problem of acid rain caused by transportation fuels (Ogunkunle & Ahmed, 2021). Neither green diesel nor biodiesel can compete with petroleum diesel in terms of energy density.

Considering that the biodiesel production process has raised critical issues such as the food-versus-fuel debate, land use change, microbial contamination in fuel tanks and so on, alternative conversion technologies are now being extensively researched. The current research focus is primarily on hydroprocessing vegetable oils and/or animal wastes to produce hydrotreated vegetable oils that can be used without blending restrictions. Catalytic hydroprocessing has also been used to create renewable diesel from residual lipids, specifically waste cooking oil. This high-quality fuel, known as "white diesel," can be used in conventional diesel vehicles, either alone or in combination with fossil diesel. White diesel is a sulphur-free, aromatic-free paraffinic fuel (Chrysikou, Dagonikou, Dimitriadis, & Bezergianni, 2019; Fernández-Villamil & Paniagua, 2018). Apart from that, vegetable oil can also be coprocessed with petroleum-derived raw materials in order to produce diesel-range hydrocarbons. By integrating liquid biomass into typical refining operations for the production of hybrid fuels, this solution enables the use of existing refinery technologies, such as catalytic hydroprocessing and equipment. In addition, hybrid diesels are regarded "clean fuels" because they are low sulphur and aromatics free (Bezergianni & Dimitriadis, 2013; Bezergianni, Dimitriadis, & Chrysikou, 2016). Table 2.8 serves to illustrates the distinction between various types of renewable diesel and fossil diesel.

Table 2.8 Comparison of various types of renewable diesel with fossil diesel (Yoon, 2009).

Type of diesel	Renewable				Fossil
Properties	Green diesel	Biodiesel	White diesel	Hybrid diesel (VGO + VO)	Petroleum diesel
Carbon	84.9	76.2	86.67	n/a	86.8
Hydrogen	15.1	12.6	14.74	n/a	13.2
Oxygen	0	11.2	0	n/a	0
Water, mg/kg	42–95	28.5–500	13	10–50	0.5
Density, g/mL	0.78	0.88	0.79	0.78–0.85	0.83–0.85
Cetane number	75–90	50–65	n/a	50–101	40–55
Cetane index	50–105	58.3	77.23	51–64	54.57
Cloud point, °C	−10	20	n/a	(−23) −20	−5
Flash point, °C	>59	100–180	116	116	54–148
Pour point, °C	(−3) −29	(−15) −16	23	(−26) −20	−21
Distillation, 90 vol.%, °C	298–342	n/a	302.6	300–332	341
Cold filter plugging point (CFPP), °C	>20	(−13) −15	20	(−24) −22	−6
Viscosity at 40°C, mm^2/s	2–4	2.9–11	3.5	2.7–5.5	1.9–4.1
Heating value, MJ/kg	42–44	≈37.2	≈49	43.3–47	≈43.1
MCRT carbon residue, wt.%	n/a	0.02–3	0.0066	85.8	n/a
Energy density, MJ/kg	44	38	n/a	n/a	43
Energy content, BTU/gal	123k	118k	n/a	n/a	129k
Induction time, hour (110°C)	>22	0.9–10.9	>22	>22	n/a
Copper strip corrosion (3 hours; 50°C)	n/a	1	1b	n/a	< 3
Sulphur content	<10 ppm	<5 ppm	1.54	3–13	<10 ppm
Colour (ASTM)	≈2	n/a	0	n/a	n/a
Ash content (wt.%)	<0.001	n/a	n/a	n/a	0.01
NOx emission	−10 to 0	+ 10	n/a	n/a	Baseline
Lubricity	Excellent	Poor	n/a	n/a	Baseline
Oxidative stability	Excellent	Poor	High	n/a	Baseline
Cold flow properties	Similar	Excellent	n/a	n/a	Baseline

Source: From Amin, A. (2019). Review of diesel production from renewable resources: Catalysis, process kinetics and technologies. Ain Shams Engineering Journal, 10*(4), 821–839. https://doi.org/10.1016/j.asej.2019.08.001; Bezergianni, S., & Dimitriadis, A. (2013). Comparison between different types of renewable diesel.* Renewable and Sustainable Energy Reviews, 21, *110–116. https://doi.org/10.1016/j.rser.2012.12.042; Douvartzides, S. L., Charisiou, N. D., Papageridis, K. N., & Goula, M. A. (2019). Green diesel: Biomass feedstocks, production technologies, catalytic research, fuel properties and performance in compression ignition internal combustion engines.* Energies, 12*(5). https://doi.org/10.3390/en12050809; and Yoon, J. J. (2009). What's the difference between biodiesel and renewable (green) diesel.* Advanced Biofuels USA, *1–14.*

2.10 Conclusion

Renewable diesel offers a promising alternative to diesel fuel. This review covered in depth the most prevalent conversion techniques for producing different types of renewable diesel. The comparison of various renewable diesels provided a concise insight and guidance on how to reflect on the choices of applying renewable diesel to fit the various applications. Renewable diesel has numerous advantages. In addition to being renewable, the renewable diesel is much cleaner than the conventional fuel and results in fewer greenhouse gas emissions. Given all these perks, the reason renewable diesel has yet to become commonly accessible is undeniably processing costs. The high processing cost to produce renewable diesel may thwart many companies and government agencies from using it. As a result, a relatively low processing cost to produce renewable diesel was expeditiously urged to eventually replace non-renewable fuels to reduce the carbon footprint and sustain energy supply. Long-term commercialisation of renewable diesel production requires government funding or support, as well as public awareness of sustainable concepts to promote the benefits of renewable diesel over conventional fuels. It is also extremely crucial to interconnect industrial and academic institutions work together on this project. With their respective expertise, it helps to accelerate the commercialisation process for renewable diesel production. With all this remarkable engagement, renewable diesel does indeed have a bright and lucrative future

References

Abed, K. A., El Morsi, A. K., Sayed, M. M., Shaib, A. A. E., & Gad, M. S. (2018). Effect of waste cooking-oil biodiesel on performance and exhaust emissions of a diesel engine. *Egyptian Journal of Petroleum, 27*(4), 985–989. Available from https://doi.org/10.1016/j.ejpe.2018.02.008.

Abukhadra, M. R., & Sayed, M. A. (2018). K + trapped kaolinite (Kaol/K +) as low cost and eco-friendly basic heterogeneous catalyst in the transesterification of commercial waste cooking oil into biodiesel. *Energy Conversion and Management, 177*(September), 468–476. Available from https://doi.org/10.1016/j.enconman.2018.09.083.

Afzal, A., Ağbulut, Ü., Soudagar, M. E. M., Razak, R. K. A., Buradi, A., & Saleel, C. A. (2021). Blends of scum oil methyl ester, alcohols, silver nanoparticles and the operating conditions affecting the diesel engine performance and emission: an optimisation study using Dragon fly algorithm. *Applied Nanoscience, 11*(9), 2415–2432. Available from https://doi.org/10.1007/s13204-021-02046-5.

Agarwal, A. K. (2007). Biofuels (alcohols and biodiesel) applications as fuels for internal combustion engines. *Progress in Energy and Combustion Science, 33* (3), 233–271. Available from https://doi.org/10.1016/j.pecs.2006.08.003.

Ahadan, N. F., Adipoetra, H., Nuryati, A., Dewi, G. P., & Nasikin, M. (2021). Comparison of pyrogallol derivative performance using methyl linoleate from sunflower oil and corn oil as biodiesel antioxidant additives. *IOP Conference Series: Earth and Environmental Science, 749*(1), 012035. Available from https://doi.org/10.1088/1755-1315/749/1/012035.

Alajtal, A. I., Elbagerma, M. A., Alajtal, A. I., Sherami, F. E., Elbagermi, M. A., Atiga, M., & Acid, E. (2018). Acid, peroxide, ester and saponification values for some vegetable oils before and after frying. *AASCIT Journal of Materials, 4*(2), 43–47.

Alherbawi, M., McKay, G., Mackey, H. R., & Al-Ansari, T. (2021). Jatropha curcas for jet biofuel production: Current status and future prospects. *Renewable and Sustainable Energy Reviews, 135*, 110396. Available from https://doi.org/10.1016/j.rser.2020.110396.

Alviso, D., Zárate, C., & Duriez, T. (2021). Modeling of vegetable oils cloud point, pour point, cetane number and iodine number from their composition using genetic programming. *Fuel, 284*(September 2020), 119026. Available from https://doi.org/10.1016/j.fuel.2020.119026.

Ameen, M., Azizan, M. T., Yusup, S., Ramli, A., & Yasir, M. (2017). Catalytic hydrodeoxygenation of triglycerides: An approach to clean diesel fuel production. *Renewable and Sustainable Energy Reviews, 80*(May), 1072–1088. Available from https://doi.org/10.1016/j.rser.2017.05.268.

Amin, A. (2019). Review of diesel production from renewable resources: Catalysis, process kinetics and technologies. *Ain Shams Engineering Journal, 10*(4), 821–839. Available from https://doi.org/10.1016/j.asej.2019.08.001.

Attaphaiboon, W., Neramittagapong, A., Neramittagapong, S., & Theerakulpisut, S. (2021). Potential of vegetable oils for producing green diesel via hydrocracking process. *Thai Environmental Engineering Journal, 35*(2), 1–11.

Bangjang, T., Kaewchada, A., & Jaree, A. (2021). Hydroprocessing of palm oil using Rh/HZSM-5 for the production of biojet fuel in a fixed bed reactor. *The Canadian Journal of Chemical Engineering, 99*(2), 435–446. Available from https://doi.org/10.1002/cjce.23877.

Bezergianni, S., & Chrysikou, L. P. (2012). Oxidative stability of waste cooking oil and white diesel upon storage at room temperature. *Bioresource Technology, 126*, 341–344. Available from https://doi.org/10.1016/j.biortech.2012.09.136.

Bezergianni, S., & Dimitriadis, A. (2013). Comparison between different types of renewable diesel. *Renewable and Sustainable Energy Reviews, 21*, 110–116. Available from https://doi.org/10.1016/j.rser.2012.12.042.

Bezergianni, S., Dimitriadis, A., & Chrysikou, L. (2016). Cost-effective and sustainable decarbonisation of aviation fuels. In *2nd ECATS conference making aviation environmentally sustainable* (pp. 88–91). Athens, Greece: ECATS.

Bezergianni, S., Dimitriadis, A., Kalogianni, A., & Pilavachi, P. A. (2010). Hydrotreating of waste cooking oil for biodiesel production. Part I: Effect of temperature on product yields and heteroatom removal. *Bioresource Technology, 101*(17), 6651–6656. Available from https://doi.org/10.1016/j.biortech.2010.03.081.

Bezergianni, S., Dimitriadis, A., & Voutetakis, S. (2011). Catalytic hydrotreating of waste cooking oil for white diesel production. In *Proceedings of the first international* congress on catalysis for biorefineries *(CatBior)*, 1–6 October 2011. Torremolinos-Málaga, Spain.

Bharti, M. K., Chalia, S., Thakur, P., Sridhara, S. N., Thakur, A., & Sharma, P. B. (2021). Nanoferrites heterogeneous catalysts for biodiesel production from soybean and canola oil: A review. *Environmental Chemistry Letters, 19*(5), 3727–3746. Available from https://doi.org/10.1007/s10311-021-01247-2.

Binhweel, F., Bahadi, M., Pyar, H., Alsaedi, A., Hossain, S., & Ahmad, M. I. (2021). A comparative review of some physicochemical properties of biodiesels synthesised from different generations of vegetative oils. *Journal of Physics: Conference Series, 1900*(1), 012009. Available from https://doi.org/10.1088/1742-6596/1900/1/012009.

Bukkarapu, K. R. (2019). Comparative study of different biodiesel–diesel blends. *International Journal of Ambient Energy, 40*(3), 295–303. Available from https://doi.org/10.1080/01430750.2017.1393775.

Chhetri, A. B., Watts, K. C., & Islam, M. R. (2008). Waste cooking oil as an alternate feedstock for biodiesel production. *Energies, 1*(1), 3–18. Available from https://doi.org/10.3390/en1010003.

Chia, S. R., Nomanbhay, S., Ong, M. Y., Shamsuddin, A. H., Bin., Chew, K. W., & Show, P. L. (2022). Renewable diesel as fossil fuel substitution in Malaysia: A review. *Fuel, 314*, 123137. Available from https://doi.org/10.1016/j.fuel.2022.123137.

Chinasamy, D. K. (2021). *An insight into biodiesel production from canola oil by homogeneously catalysed transesterification reaction in the presence of ethanol.* University of KwaZulu-Natal.

Chrysikou, L. P., Dagonikou, V., Dimitriadis, A., & Bezergianni, S. (2019). Waste cooking oils exploitation targeting EU 2020 diesel fuel production: Environmental and economic benefits. *Journal of Cleaner Production, 219*, 566–575. Available from https://doi.org/10.1016/j.jclepro.2019.01.211.

Doğan, T. H., & Naktiyok, J. (2022). Thermal behaviours at low and high temperature of biodiesels produced from beef tallow and corn oil. *Journal of Thermal Analysis and Calorimetry*, 1–11. Available from https://doi.org/10.1007/s10973-021-11160-z.

dos Santos, R. G., & Alencar, A. C. (2020). Biomass-derived syngas production via gasification process and its catalytic conversion into fuels by Fischer Tropsch synthesis: A review. *International Journal of Hydrogen Energy, 45*(36), 18114–18132. Available from https://doi.org/10.1016/j.ijhydene.2019.07.133.

Doustdar, O., Wyszynski, M. L., Mahmoudi, H., & Tsolakis, A. (2016). Enhancing the properties of Fischer-Tropsch fuel produced from syngas over Co/SiO2 catalyst: Lubricity and calorific value. *IOP Conference Series: Materials Science and Engineering, 148*(1). Available from https://doi.org/10.1088/1757-899X/148/1/012092.

Douvartzides, S. L., Charisiou, N. D., Papageridis, K. N., & Goula, M. A. (2019). Green diesel: Biomass feedstocks, production technologies, catalytic research, fuel properties and performance in compression ignition internal combustion engines. *Energies, 12*(5). Available from https://doi.org/10.3390/en12050809.

Dugala, N. S., Goindi, G. S., & Sharma, A. (2021). Evaluation of physicochemical characteristics of mahua (Madhuca indica) and jatropha (Jatropha curcas) dual biodiesel blends with diesel. *Journal of King Saud University - Engineering Sciences, 33*(6), 424–436. Available from https://doi.org/10.1016/j.jksues.2020.05.006.

Dwivedi, G., & Sharma,, M. P. (2013). Performance evaluation of diesel engine using biodiesel from Pongamia oil. *International Journal of Renewable Energy Research, 3*(2), 325–330.

Ebenaza, G., & Vinoth, E. (2015). Biodiesel production from waste cooking oil. *International Journal of Students' Research In Technology & Management, 3* (8), 448–450. Available from https://doi.org/10.18510/ijsrtm.2015.383.

Eke-Ejiofor, J., Beleya, E. A., & Allen, J. E. (2021). Effect of variety on the quality parameters of crude soybean oil. *American Journal of Food Science and Technology, 9*(3), 69–75. Available from https://doi.org/10.12691/ajfst-9-3-1.

Ekop, S., Etuk, B., & Eddy, N. (2007). Effect of some local additives on the chemical constituent of palm oil. *Journal Applied Sciences and Environmental Management (JASEM), 11*(1), 85–89.

Ewunie, G. A., Morken, J., Lekang, O. I., & Yigezu, Z. D. (2021). Factors affecting the potential of Jatropha curcas for sustainable biodiesel production: A critical review. *Renewable and Sustainable Energy Reviews, 137*, 110500. Available from https://doi.org/10.1016/j.rser.2020.110500.

Faizal, M., & Ateeb, S. (2018). Energy, economic and environmental impact of palm oil biodiesel in Malaysia. *Journal of Mechanical Engineering Research and Developments, 41*(3), 24–26. Available from https://doi.org/10.26480/jmerd.03.2018.24.26.

Farid, M. A. A., Roslan, A. M., Hassan, M. A., Hasan, M. Y., Othman, M. R., & Shirai, Y. (2020). Net energy and techno-economic assessment of biodiesel production from waste cooking oil using a semi-industrial plant: A Malaysia perspective. *Sustainable Energy Technologies and Assessments, 39*(September 2019), 100700. Available from https://doi.org/10.1016/j.seta.2020.100700.

Farooqui, S. A., Kumar, R., Sinha, A. K., & Ray, A. (2022). Green diesel production by hydroprocessing technology. *Green diesel: An alternative to biodiesel and petrodiesel* (pp. 109–148). Singapore: Springer. Available from https://doi.org/10.1007/978-981-19-2235-0_4.

Fayad, M. A., Chaichan, M. T., & Dhahad, H. A. (2021). Engine performance and PM concentrations from the combustion of Iraqi sunflower oil biodiesel under variable diesel engine operating conditions. *Journal of Physics: Conference Series, 1973*, 012051. Available from https://doi.org/10.1088/1742-6596/1973/1/012051.

Fernández-Villamil, J. M., & Paniagua, A. H. D. M. (2018). Preliminary design of the green diesel production process by hydrotreatment of vegetable oils. *Eurecha*, 15.

Gangolu, N. R., Chunchu, B. R. K., Yallamati, A., & Gopidesi, R. K. (2021). Assessment of diesel engine characteristics by using soybean oil as a biofuel. *Environment, Development and Sustainability*, 1–14. Available from https://doi.org/10.1007/s10668-021-01645-0.

Ge, J. C., Yoon, S. K., & Choi, N. J. (2017). Using canola oil biodiesel as an alternative fuel in diesel engines: A review. *Applied Sciences (Switzerland), 7* (9). Available from https://doi.org/10.3390/app7090881.

Gholami, A., Pourfayaz, F., & Maleki, A. (2021). Techno-economic assessment of biodiesel production from canola oil through ultrasonic cavitation. *Energy Reports, 7*, 266–277. Available from https://doi.org/10.1016/j.egyr.2020.12.022.

Gholizadeh, M., Hu, X., & Liu, Q. (2019). A mini review of the specialties of the bio-oils produced from pyrolysis of 20 different biomasses. *Renewable and Sustainable Energy Reviews, 114*(August), 109313. Available from https://doi.org/10.1016/j.rser.2019.109313.

Goga, G., Chauhan, B. S., Mahla, S. K., Cho, H. M., Dhir, A., & Lim, H. C. (2018). Properties and characteristics of various materials used as biofuels: A review. *Materials Today: Proceedings, 5*(14), 28438–28445. Available from https://doi.org/10.1016/j.matpr.2018.10.130.

Gopinath, A., Puhan, S., & Nagarajan, G. (2010). Effect of unsaturated fatty acid esters of biodiesel fuels on combustion, performance and emission characteristics of a DI diesel engine. *International Journal of Energy and Environment, 1*(3), 411–430.

Guharaja, S., DhakshinaMoorthy, S., Hasan, Z. I., Arun, B., Ahamed, J. I., & Azarudheen, J. (2016). Biodiesel production from Mahua (Madhuca indica). *International Journal of Nano Corrosion Science and Engineering, 34*.

Hachemane, K. (2022). Study of the application of choline hydroxide as a catalyst in biodiesel production. University of SAIDA.

Hartikka, T., Kiiski, U., Kuronen, M., & Mikkonen, S. (2013). Diesel fuel oxidation stability: A comparative study. *SAE Technical Papers, 11*. Available from https://doi.org/10.4271/2013-01-2678.

Hazmi, B., Rashid, U., Kawi, S., Mokhtar, W. N. A. W., Yaw, T. C. S., Moser, B. R., & Alsalme, A. (2022). Palm fatty acid distillate esterification using synthesised heterogeneous sulfonated carbon catalyst from plastic waste: Characterisation, catalytic efficacy and stability, and fuel properties. *Process Safety and Environmental Protection, 162*, 1139–1151. Available from https://doi.org/10.1016/j.psep.2022.05.001.

Hazmi, B., Rashid, U., Kawi, S., Nur, W., Wan, A., & Ngamcharussrivichai, C. (2022). A heterogeneous bifunctional carbon nanocatalyst from plastic waste to efficiently catalyze waste cooking oil into biodiesel. *Catalysts, 12*(8), 874. Available from https://doi.org/10.3390/catal12080874.

Ibrahim, S. F., Asikin-Mijan, N., Ibrahim, M. L., Abdulkareem-Alsultan, G., Izham, S. M., & Taufiq-Yap, Y. H. (2020). Sulfonated functionalisation of carbon derived corncob residue via hydrothermal synthesis route for esterification of palm fatty acid distillate. *Energy Conversion and Management, 210*(March). Available from https://doi.org/10.1016/j.enconman.2020.112698.

Ibrahim, S. M. A., Abed, K. A., Gad, M. S., & Abu Hashish, H. M. (2020). Comparison of different methods for producing bio oil from Egyptian jatropha seeds. *Biofuels, 11*(6), 643–654. Available from https://doi.org/10.1080/17597269.2017.1387748.

Ilkiliç, C., & Yücesu, H. S. (2005). Investigation of the effect of sunflower oil methyl esther on the performance of a diesel engine. *Energy Sources, 27*(13), 1225–1234. Available from https://doi.org/10.1080/009083190519311.

Kelly, S. (2021). U.S. renewable fuels market could face feedstock deficit. Reuters.

Khanal, A., & Shah, A. (2021). Oilseeds to biodiesel and renewable jet fuel: an overview of feedstock production, logistics, and conversion. *Biofuels, Bioproducts and Biorefining, 15*(3), 913–930. Available from https://doi.org/10.1002/bbb.2198.

Kittisupakorn, P., Sae-ueng, S., & Suwatthikul, A. (2016). Optimisation of energy consumption in a hydrotreating process for green diesel production from palm oil. *Computer Aided Chemical Engineering, 38*, 751–756.

Knothe, G. (2010). Biodiesel and renewable diesel: A comparison. *Progress in Energy and Combustion Science, 36*(3), 364–373.

Kordouli, E., Kordulis, C., Lycourghiotis, A., Cole, R., Vasudevan, P. T., Pawelec, B., & Fierro, J. L. G. (2017). HDO activity of carbon-supported Rh, Ni and Mo-Ni catalysts. *Molecular Catalysis, 441*, 209–220. Available from https://doi.org/10.1016/j.mcat.2017.08.013.

Kordouli, E., Sygellou, L., Kordulis, C., Bourikas, K., & Lycourghiotis, A. (2017). Probing the synergistic ratio of the NiMo/Γ-Al_2O_3 reduced catalysts for the transformation of natural triglycerides into green diesel. *Applied Catalysis B: Environmental, 209*, 12–22. Available from https://doi.org/10.1016/j.apcatb.2017.02.045.

Koul, R., Kumar, N., & Singh, R. C. (2021). A review on the production and physicochemical properties of renewable diesel and its comparison with biodiesel. *Energy Sources, Part A: Recovery, Utilisation and Environmental Effects, 43*(18), 2235–2255. Available from https://doi.org/10.1080/15567036.2019.1646355.

Koul, R., Kumar, N., & Singh, R. C. (2022). Comparative analysis of renewable diesel and biodiesel produced from jatropha oil. *Environmental Progress & Sustainable Energy*, 13832.

Kumar, A. (2014). Coconut oil as an hybrid fuel for diesel engine. *International Journal of Engineering Research & Technology, 3*(11), 1082–1086.

Kumar, N., & Raheman, H. (2021). Combustion, performance, and emission characteristics of diesel engine with nanocera added water emulsified Mahua biodiesel blend. *Environmental Progress and Sustainable Energy, 40*(4), e13572. Available from https://doi.org/10.1002/ep.13572.

Kumar, O. M. C., & Simhadri, K. (2021). Exploration of mahua oil (Madhucaindica seed oil) biodiesel as a fuel in compression ignition engine: A review. *NVEO - Natural Volatiles & Essential Oils, 8*(4), 11609–11627.

Li, P., Shi, X., Wang, X., Song, J., Fang, S., Bai, J., … Pang, S. (2021). Bio-oil from biomass fast pyrolysis: Yields, related properties and energy consumption analysis of the pyrolysis system. *Journal of Cleaner Production, 328*(October), 129613. Available from https://doi.org/10.1016/j.jclepro.2021.129613.

Linganiso, E. C., Tlhaole, B., Magagula, L. P., Dziike, S., Linganiso, L. Z., Motaung, T. E., … Tetana, Z. N. (2022). Biodiesel production from waste oils: A South African outlook. *Sustainability (Switzerland), 14*, 1983. Available from https://doi.org/10.3390/su14041983.

Liu, J., Lei, J., He, J., Deng, L., Wang, L., Fan, K., & Rong, L. (2015). Hydroprocessing of jatropha oil for production of green diesel over non-sulfided Ni-PTA/Al2O3 catalyst. *Scientific Reports, 5*, 1–13. Available from https://doi.org/10.1038/srep11327.

Liu, X., Driskell, R. R., & Engelhardt, J. F. (2006). Stem cells in the lung. *Methods in Enzymology, 419*(06), 285–321. Available from https://doi.org/10.1016/S0076-6879(06)19012-6.

Ma'arof, N. A. N. B., Hindryawati, N., Khazaai, S. N. M., Bhuyar, P., Rahim, M. H. A., & Maniam, G. P. (2021). Exploitation of cost-effective renewable heterogeneous base catalyst from banana (*Musa paradisiaca*) peel for effective methyl ester production from soybean oil. *Applied Nanoscience*, 1–12. Available from https://doi.org/10.1007/s13204-021-01926-0.

Mehta, P. S., & Anand, K. (2009). Estimation of a lower heating value of vegetable oil and biodiesel fuel. *Energy and Fuels, 23*(8), 3893–3898. Available from https://doi.org/10.1021/ef900196r.

Mekhilef, S., Siga, S., & Saidur, R. (2011). A review on palm oil biodiesel as a source of renewable fuel. *Renewable and Sustainable Energy Reviews, 15*(4), 1937–1949. Available from https://doi.org/10.1016/J.RSER.2010.12.012.

Mikkonen, S. (2008). Second-generation renewable diesel offers advantages. *Hydrocarbon Processing, 87*(2), 63–66.

Montefrio, M. J., Xinwen, T., & Obbard, J. P. (2010). Recovery and pre-treatment of fats, oil and grease from grease interceptors for biodiesel production. *Applied Energy, 87*(10), 3155–3161. Available from https://doi.org/10.1016/j.apenergy.2010.04.011.

Moreira, J., Rezende, D., & Pasa, V. (2020). Deoxygenation of Macauba acid oil over Co-based catalyst supported on activated biochar from Macauba endocarp: A potential and sustainable route for green diesel and biokerosene production. *Fuel, 269*(January), 117253. Available from https://doi.org/10.1016/j.fuel.2020.117253.

Morrissey, K. G., Thoma, G., & López, D. E. (2021). Life cycle impact assessment of biofuels derived from sweet sorghum in the U.S. *Biotechnology for Biofuels, 14*(1), 1–11. Available from https://doi.org/10.1186/s13068-021-02009-6.

Munyentwali, A., Li, H., & Yang, Q. (2022). Review of advances in bifunctional solid acid/base catalysts for sustainable biodiesel production. *Applied Catalysis A: General, 633*(November 2021). Available from https://doi.org/10.1016/j.apcata.2022.118525.

Muruganantham, P., Pandiyan, P., & Sathyamurthy, R. (2021). Analysis on performance and emission characteristics of corn oil methyl ester blended with diesel and cerium oxide nanoparticle. *Case Studies in Thermal Engineering, 26*, 101077. Available from https://doi.org/10.1016/j.csite.2021.101077.

Neuana, N. F., de Sousa Barboza, J. C., dos Santos, E. P., & da Silva, M. L. C. P. (2021). A novel application of *Mangifera indica* L. and *Eugenia uniflora* L. extracts as antioxidants to control biodiesel oxidation stability. *Environmental Progress and Sustainable Energy, 40*(3). Available from https://doi.org/10.1002/ep.13540.

Ogunkunle, O., & Ahmed, N. A. (2021). Overview of biodiesel combustion in mitigating the adverse impacts of engine emissions on the sustainable human–environment scenario. *Sustainability (Switzerland), 13*(10). Available from https://doi.org/10.3390/su13105465.

Padhi, S. K., & Singh, R. K. (2010). Optimisation of esterification and transesterification of Mahua (*Madhuca indica*) oil for production of biodiesel. *Journal of Chemical and Pharmaceutical Research, 2*(5), 599–608.

Patel, M., & Kumar, A. (2016). Production of renewable diesel through the hydroprocessing of lignocellulosic biomass-derived bio-oil: A review. *Renewable and Sustainable Energy Reviews, 58*, 1293–1307. Available from https://doi.org/10.1016/j.rser.2015.12.146.

Peter, A. S., Alias, M. P., Iype, M. P., Jolly, J., Sankar, V., Joseph Babu, K., & Baby, D. K. (2020). Optimisation of biodiesel production by transesterification of palm oil and evaluation of biodiesel quality. *Materials Today: Proceedings, 42*, 1002–1007. Available from https://doi.org/10.1016/j.matpr.2020.11.995.

Pioch, D., & Vaitilingom, G. (2005). Palm oil and derivatives: Fuels or potential fuels? *Oléagineux, Corps Gras, Lipides, 12*(2), 161–169. Available from https://doi.org/10.1051/ocl.2005.0161.

Puhan, S., Nagarajan, G., Vedaraman, N., & Ramabramhmam, B. V. (2007). Mahua oil (madhuca indica oil) derivatives as a renewable fuel for diesel engine systems in India: A performance and emissions comparative study. *International Journal of Green Energy, 4*(1), 89–104. Available from https://doi.org/10.1080/15325000601015627.

Rama Krishna Reddy, E., Subbalakshmi, Y., Dhana Raju, V., Appa Rao, K., Harun Kumar, M., Rami Reddy, S., & Tharun Sai, P. (2022). Assessment of performance, combustion and emission characteristics of the diesel engine powered with corn biodiesel blends. *International Journal of Ambient Energy, 43*(1), 435–443. Available from https://doi.org/10.1080/01430750.2019.1636867.

Rashid, U., Ibrahim, M., Yasin, S., Yunus, R., Taufiq-Yap, Y. H., & Knothe, G. (2013). Biodiesel from Citrus reticulata (mandarin orange) seed oil, a potential non-food feedstock. *Industrial Crops and Products, 45*, 355–359. Available from https://doi.org/10.1016/j.indcrop.2012.12.039.

Rathore, Y., & Ramchandani, D. (2016). Mahua biodiesel: An alternative to conventional diesel fuel. *IOSR Journal of Mechanical and Civil Engineering, 3*, 42–51. Available from https://doi.org/10.9790/1684-15010030342-51.

Rauch, R., Kiennemann, A., & Sauciuc, A. (2013). Fischer–Tropsch synthesis to biofuels (BtL process). *The role of catalysis for the sustainable production of bio-fuels and bio-chemicals*. Elsevier B.V. Available from https://doi.org/10.1016/B978-0-444-56330-9.00012-7.

Ruangudomsakul, M., Osakoo, N., Wittayakun, J., Keawkumay, C., Butburee, T., Youngjan, S., … Khemthong, P. (2021). Hydrodeoxygenation of palm oil to green diesel products on mixed-phase nickel phosphides. *Molecular Catalysis*, 111422. Available from https://doi.org/10.1016/j.mcat.2021.111422.

Safa-Gamal, M., Asikin-Mijan, N., Arumugam, M., Khalit, W. N. A. W., Azreena, N., Hafez, F. S., & Taufiq-Yap, Y. H. (2021). Catalytic deoxygenation by H2-free single-step conversion of free fatty acid feedstock over a Co-Ag carbon-based catalyst for green diesel production. *Journal of Analytical and Applied Pyrolysis*, *160*(September), 105334. Available from https://doi.org/10.1016/j.jaap.2021.105334.

Sahoo, P. K., & Das, L. M. (2009). Process optimisation for biodiesel production from Jatropha, Karanja and Polanga oils. *Fuel*, *88*(9), 1588–1594. Available from https://doi.org/10.1016/j.fuel.2009.02.016.

Saydut, A., Kafadar, A. B., Tonbul, Y., Kaya, C., Aydin, F., & Hamamci, C. (2010). Comparison of the biodiesel quality produced from refined sunflower (*Helianthus annuus* L.) oil and waste cooking oil. *Energy Exploration and Exploitation*, *28*(6), 499–512. Available from https://doi.org/10.1260/0144-5987.28.6.499.

Seffati, K., Honarvar, B., Esmaeili, H., & Esfandiari, N. (2019). Enhanced biodiesel production from chicken fat using $CaO/CuFe_2O_4$ nanocatalyst and its combination with diesel to improve fuel properties. *Fuel*, *235*(August 2018), 1238–1244. Available from https://doi.org/10.1016/j.fuel.2018.08.118.

Shaban, M., Hosny, R., Rabie, A. M., Shim, J. J., Ahmed, S. A., Betiha, M. A., & Negm, N. A. (2020). Zinc aluminate nanoparticles: Preparation, characterisation and application as efficient and economic catalyst in transformation of waste cooking oil into biodiesel. *Journal of Molecular Liquids*, *302*, 112377. Available from https://doi.org/10.1016/j.molliq.2019.112377.

Silitonga, A. S., Masjuki, H. H., Mahlia, T. M. I., Ong, H. C., Atabani, A. E., & Chong, W. T. (2013). A global comparative review of biodiesel production from Jatropha curcas using different homogeneous acid and alkaline catalysts: Study of physical and chemical properties. *Renewable and Sustainable Energy Reviews*, *24*, 514–533. Available from https://doi.org/10.1016/j.rser.2013.03.044.

Singh, B., Guldhe, A., Rawat, I., & Bux, F. (2014). Towards a sustainable approach for development of biodiesel from plant and microalgae. *Renewable and Sustainable Energy Reviews*, *29*, 216–245. Available from https://doi.org/10.1016/j.rser.2013.08.067.

Singh, D., Sharma, D., Soni, S. L., Inda, C. S., Sharma, S., Sharma, P. K., & Jhalani, A. (2021). A comprehensive review of physicochemical properties, production process, performance and emissions characteristics of 2nd generation biodiesel feedstock: Jatropha curcas. *Fuel*, *285*, 119110. Available from https://doi.org/10.1016/j.fuel.2020.119110.

Singh, D., Sharma, D., Soni, S. L., Sharma, S., Kumar Sharma, P., & Jhalani, A. (2020). A review on feedstocks, production processes, and yield for different generations of biodiesel. *Fuel*, *262*(July 2019), 116553. Available from https://doi.org/10.1016/j.fuel.2019.116553.

Soltani, S., Rashid, U., Nehdi, I. A., Al-Resayes, S. I., & Al-Muhtaseb, A. H. (2017). Sulfonated mesoporous zinc aluminate catalyst for biodiesel production from high free fatty acid feedstock using microwave heating system. *Journal of the*

Taiwan Institute of Chemical Engineers, 70, 219–228. Available from https://doi.org/10.1016/j.jtice.2016.10.054.

Stanciu, I. (2018). Viscosity index for soybean oil. *MAGNT Research Report, 5*(3), 447–449.

Stanciu, I. (2020). Viscosity index for oil used as biodegradable lubricant. *Indian Journal of Science and Technology, 13*(3), 352–359. Available from https://doi.org/10.17485/ijst/2020/v13i03/147759.

Subroto, E. (2020). Characteristics, purification, and the recent applications of soybean oil in fat-based food products: A review. *International Journal of Emerging Trends in Engineering Research, 8*(7), 3003–3011. Available from https://doi.org/10.30534/ijeter/2020/20872020.

Tahir, A. R., Lapp, H. M., & Buchanan, L. C. (1982). Sunflower oil as a fuel for compression ignition engines. *American Society of Agricultural Engineers, 4* (82), 82–91.

Teo, S. H., Islam, A., Chan, E. S., Thomas Choong, S. Y., Alharthi, N. H., Taufiq-Yap, Y. H., & Awual, M. R. (2019). Efficient biodiesel production from Jatropha curcus using $CaSO_4/Fe_2O_3$-SiO_2 core-shell magnetic nanoparticles. *Journal of Cleaner Production, 208*, 816–826. Available from https://doi.org/10.1016/j.jclepro.2018.10.107.

Teresa L. Alleman, R. L. M. (2016). *Biodiesel handling and use guide* (5th ed.). Golden, CO: National Renewable Energy Laboratory (NREL).

Tian, K., Tai, K., Chua, B. J. W., & Li, Z. (2017). Directed evolution of Thermomyces lanuginosus lipase to enhance methanol tolerance for efficient production of biodiesel from waste grease. *Bioresource Technology, 245*, 1491–1497. Available from https://doi.org/10.1016/j.biortech.2017.05.108.

Tucker, C. L., Bordoloi, A., & van Steen, E. (2021). Novel single pass biogas-to-diesel process using a Fischer–Tropsch catalyst designed for high conversion. *Sustainable Energy and Fuels, 5*(22), 5717–5732. Available from https://doi.org/10.1039/d1se01299a.

Tuli, D., & Kasture, S. (2022). Biodiesel and green diesel. *Advanced Biofuel Technologies*, 119–133. Available from https://doi.org/10.1016/b978-0-323-88427-3.00010-6.

Tutunea, D., Dumitru, I., Racila, L., Otat, O., Matei, L., & Geonea, I. (2019). *Characterisation of sunflower oil biodiesel as alternative for diesel fuel.* Springer International Publishing. Available from https://doi.org/10.1007/978-3-319-94409-8_21.

Tyagi, U., Aslam, M., & Sarma, A. K. (2022). Characterisation of green diesel: Existing standards and beyond. *Green diesel: An alternative to biodiesel and petrodiesel* (pp. 249–263). Singapore: Springer. Available from https://doi.org/10.1007/978-981-19-2235-0_8.

Vignesh, P., Pradeep Kumar, A. R., Ganesh, N. S., Jayaseelan, V., & Sudhakar, K. (2021). Biodiesel and green diesel generation: An overview. *Oil and Gas Science and Technology, 76*(6), 1–15. Available from https://doi.org/10.2516/ogst/2020088.

Vlnieska, V., Muniz, A. S., Oliveira, A. R. D. S., César-Oliveira, M. A. F., & Kunka, D. (2022). Oligocat: Oligoesters as pseudo-homogenous catalysts for biodiesel synthesis. *Polymers, 14*(1), 210. Available from https://doi.org/10.3390/polym14010210.

Xu, X., Li, Z., Tu, R., Sun, Y., & Jiang, E. (2018). Hydrogen from rice husk pyrolysis volatiles via non-noble Ni-Fe catalysts supported on five differently treated rice husk pyrolysis carbon supports. *ACS Sustainable Chemistry and Engineering, 6*(7), 8325–8339. Available from https://doi.org/10.1021/acssuschemeng.8b00343.

Yadav, A., Singh, Y., Singh, S., & Negi, P. (2021). Sustainability of vegetable oil based bio-diesel as dielectric fluid during EDM process – a review. *Materials Today: Proceedings, 46*, 11155–11158. Available from https://doi.org/10.1016/j.matpr.2021.01.967.

Yadav, M., & Sharma, Y. C. (2019). Transesterification of used vegetable oil using $BaAl_2O_4$ spinel as heterogeneous base catalyst. *Energy Conversion and Management, 198*(July), 111795. Available from https://doi.org/10.1016/j.enconman.2019.111795.

Yaqoob, H., Teoh, Y. H., Sher, F., Farooq, M. U., Jamil, M. A., Kausar, Z., … Rehman, A. U. (2021). Potential of waste cooking oil biodiesel as renewable fuel in combustion engines: A review. *Energies, 14*(9). Available from https://doi.org/10.3390/en14092565.

Yoon, J. J. (2009). What's the difference between biodiesel and renewable (green) diesel. *Advanced Biofuels USA*, 1–14.

Zacher, A. H., Olarte, M. V., Santosa, D. M., Elliott, D. C., & Jones, S. B. (2014). A review and perspective of recent bio-oil hydrotreating research. *Green Chemistry, 16*(2), 491–515. Available from https://doi.org/10.1039/c3gc41382a.

Zhao, X., Wei, L., Cheng, S., & Julson, J. (2017). Review of heterogeneous catalysts for catalytically upgrading vegetable oils into hydrocarbon biofuels. *Catalysts, 7*(3). Available from https://doi.org/10.3390/catal7030083.

Further reading

Changmai, B., Rano, R., Vanlalveni, C., & Rokhum, L. (2021). A novel *Citrus sinensis* peel ash coated magnetic nanoparticles as an easily recoverable solid catalyst for biodiesel production. *Fuel, 286*(P2), 119447. Available from https://doi.org/10.1016/j.fuel.2020.119447.

Kumar, D., Das, T., Giri, B. S., & Verma, B. (2020). Preparation and characterization of novel hybrid bio-support material immobilized from *Pseudomonas cepacia* lipase and its application to enhance biodiesel production. *Renewable Energy, 147*, 11–24. Available from https://doi.org/10.1016/j.renene.2019.08.110.

3

Life cycle assessment of renewable diesel production

Rachael J. Barla, Abhishek Anand, Smita Raghuvanshi and Suresh Gupta

Department of Chemical Engineering, Birla Institute of Technology and Science (BITS), Pilani, Rajasthan, India

3.1 Introduction

The sustainable and green approaches attract society's attention to preserve the environment for future generations. The higher energy consumption worldwide has been an issue as the primary energy feedstock as non-renewable sources is depleting rapidly due to population growth. Fossil fuel combustion and intensive anthropogenic activities have significantly increased carbon dioxide (CO_2) to a concentration of 419.13 ppm, causing global warming and climate change. Renewable fuels have received increasing attention from governments, industries and academics to reduce transportation sector pollution and greenhouse gas (GHG) emissions (Centi & Santen, 2008; Zhang et al., 2018). The global liquid fuel demand is projected to increase by more than 20% between 2018 and 2050 (U.S. Energy Information Administration, 2019). Over the past decade, alternative liquid fuels derived from biomass have gained considerable attention due to their potential for lower life cycle GHG emissions and the potential to diversify the energy supply (Chen et al., 2018; Soam & Hillman, 2019). One of the potential alternatives to decrease fossil fuel dependency and reduce GHG emissions and energy crisis is biofuel (Yee, Tan, Abdullah, & Lee, 2009). The worldwide biofuel production is expected to reach 190 billion litres/year by 2024, a 25% increase from 2018 (IEA, 2018).

Biodiesel is a biofuel similar to conventional or "fossil" diesel (Leung, Wu, & Leung, 2010), produced from raw vegetable oil, animal oils/fats, tallow oil and waste cooking oil (WCO). Biodiesel has significant environmental advantages and can potentially be a "carbon neutral" fuel. Biodiesel is attracting increasing attention worldwide as a blending component or a direct replacement for

Renewable Diesel. DOI: https://doi.org/10.1016/B978-0-323-91153-5.00003-0

diesel fuel in vehicle engines (Demirbas, 2009). Fatty acid methyl ester (FAME) "biodiesel" is the most common biofuel considered a first-generation biofuel. The primary feedstock for FAME production is vegetable oil from seeds of various crops such as sunflower, rapeseed, soy, Jatropha, etc., raising economic and social implications due to the associated "food versus fuel" debate and problematic glycerine disposal (Bezergianni & Dimitriadis, 2013). Biodiesel's low thermal and oxidation stability due to its unsaturated carbon–carbon bonding, acidity, oxygenated components and low blending ratio with petroleum diesel prompted the development of renewable or green diesel. It has a molecular structure similar to petroleum diesel and can be blended with the existing liquid fuel infrastructure (Chakraborty, Gupta, & Chowdhury, 2014; Kaewmeesri, Srifa, Itthibenchapong, & Faungnawakij, 2015).

As a result, second-generation biofuel technologies have been developed to overcome the limitations of first-generation biofuel production (Naik, Goud, Rout, & Dalai, 2010). Recently, several technologies have been developed by refinery operators and technology development companies that aim to process bio-feed stocks such as vegetable oils and animal fats into renewable diesel (also called hydro-treated vegetable oil and sometimes green diesel). Renewable diesel is a mixture of straight-chain and branched paraffin with specific carbon numbers of C_{15}–C_{18} and is chemically similar to petroleum diesel (Neste Corporation, 2015). Thus renewable diesel can be used as a drop-in fuel in engines that run on conventional diesel fuel (Zhang & Seddon, 2018).

3.1.1 Description of renewable diesel

Renewable diesel, also known as green diesel, is a diesel substitute of renewable origin (derived from vegetable oils and fats). Renewable diesel should not be confused with biodiesel (FAME) as it is produced via catalytic hydroprocessing of vegetable oils and fats and not by the transesterification reaction (Lapuerta, Armas, Hernández, & Tsolakis, 2010).

Catalytic hydroprocessing is a standard refinery process aiming to increase the hydrogen-to-carbon ratio, decrease the concentration of heteroatoms and metals and reduce the boiling point of petroleum fractions. Catalytic hydroprocessing of vegetable oil focuses on producing a high-quality biodiesel product compatible with the existing diesel fuel infrastructure. The green diesel technology comprises two steps, the catalytic hydrotreatment step, which will make normal paraffin, and the catalytic isomerisation step, leading to a mixture of *n*- and iso-paraffin (Krár, Kovács, Kalló, & Hancsók, 2010).

Renewable diesel consists mainly of paraffin and is free of aromatics, oxygen and sulphur. As a result, the paraffinic fuel has a higher cetane number and heating value than FAME. Another advantage is that the catalytic hydroprocessing leaves no by-products (propane is a common by-product of the process, plz revise the statement), unlike FAME and is accompanied by glycerine. Furthermore, as hydroprocessing includes desulphurisation reaction, renewable diesel is a low sulphur fuel (<10 ppm) with very low GHG emissions (Simacek, Kubička, Šebor, & Pospíšil, 2010). Renewable diesel can be produced from several types of vegetable oil without compromising fuel quality. However, rapeseed and palm oil are the most commonly used.

The cetane number of renewable diesel ranges between 80 and 99, much higher than the diesel standards, rendering it a competitive diesel substitute. The density range of green diesel is 0.77–0.83 g/mL, which also meets the diesel–biodiesel standard (Guzman, Torres, Prada, & Nuñez, 2010; Huber, O'Connor, & Corma, 2007; Landau, Herskowitz, Givoni, Laichter, & Yitzhaki, 1998). The net heating value is between 42 and 44 MJ/kg and is almost similar to that of conventional diesel (Aatola, Larmi, Sarjovaara, & Mikkonen, 2009; Kalnes, Marker, Shonnard, & Koers, 2008), while its low aromatic content (<0.1 wt.%) leads to cleaner combustion according to Aatola et al. (2009). Regarding its cold flow properties, its pour point ranges from −1°C to 29°C, while its cloud point ranges from −25°C to 30°C. Renewable diesel has a flashpoint of 68°C–120°C; thus, it is safe for handling and storage.

Renewable diesel is a new biofuel currently produced on an industrial scale. The first commercial-scale hydroprocessing plant with 170,000 tonnes/year was started in the summer of 2007 at Neste Oil's Porvoo oil refinery in Finland. This technology is branded as "NExBTL" (Neste Oil). For sustainable industrial development, a rigorous assessment of the environmental impacts of renewable diesel production pathways is crucial. Hence, life cycle analysis has been widely used over the past few decades to identify viable and sustainable bioenergy solutions (McManus et al., 2015; Wang, Huo, & Arora, 2011). Within this context, life cycle assessment (LCA) is a powerful method to analyse the environmental performance of products and services.

3.2 The approaches for renewable diesel production

Many alternative energy sources are required to meet the high energy demand and the effect of climate change. Biofuels such as renewable diesel and biodiesel from renewable biomass have become a sustainable alternative to fossil fuels. Developing novel, renewable and environmentally friendly resources has become a need of the hour for sustenance. Renewable diesel is also called hydro-processed vegetable oil, green diesel, hydro-treated vegetable oil or hydrogenation-derived renewable diesel (Canter, Davis, Urgun-Demirtas, & Frank, 2014). Renewable diesel has several vantages compared to biodiesel, thus making it superior. Renewable diesel has good stability, needs less investment, offers excellent pour point and flexibility in using different feedstocks (Zandonai, Yassue-Cordeiro, Castellã-Pergher, Scaliante, & Fernandes-Machado, 2016).

3.2.1 Optimisation of raw material conditions

Renewable diesel production is carried out by hydrotreatment of various naturally occurring oils and fats. Renewable diesel is chemically comparable to diesel produced by treating fossil fuels with minor modifications. Renewable diesel production depends on a similar feedstock base as biodiesel—vegetable oils, animal fats and waste oil. The primary oils and fats utilised as feedstocks are soy oil, canola oil, distiller corn oil (DCO), used cooking oil (UCO) and animal fats (Bacon, 1955). Jatropha oil is a significant feedstock for the production of renewable diesel. There are significant advantages involved with using Jatropha as it is a perennial plant, and the oil is non-edible and can be grown in arid, semi-arid and drought regions (Vibhanshu et al., 2013). Renewable diesel blends are the recent trend being researched in this field, which can replace diesel fuels. The ternary blend of specific proportions can be used for commercial applications with existing pipeline infrastructure. The blend (diesel + renewable diesel + ethanol) has renewable diesel percentages varying from 20% to 40%, ethanol from 5% to 15%, and the rest is balanced with diesel. The maximum percentage of ethanol in the blends can be 15% due to its immiscibility beyond 15% by volume (Kumar, Koul, & Singh, 2022).

Butyric acid is a capable chemical precursor for producing renewable diesel, and it can be biologically obtained from lignocellulosic sugars. Butyric acid is made either by the oxidation of butyraldehyde or by anaerobic fermentation from lignocellulose. The selling price of butyric acid from biomass is 55% of that of butyric acid from petroleum, thus promoting a decrease in renewable fuel production (Chia et al., 2022; Salvachua et al., 2021) studied the non-edible oil available in different countries and their use for renewable diesel production. Indonesia (Jatropha and Palm oil), East Indies, Philippines and India

(Karanja oil), Malaysia (Palm oil) and Ukraine and Russia (Sunflower oil) produce the specific oil in abundance. They reported the conversion of >80% of these oil into renewable diesel under optimised conditions, with selectivity to $C_{15}-C_{18}$ carbon chain length.

Another essential resource category is the wet waste feedstocks such as animal manure and fats, oils and greases (FOG). Depending on the nature of the wet waste resources, regular waste management practices are required to collect, store, treat and dispose of the waste. To avoid the added cost of waste management and disposal, conventional waste management practices can be changed to using waste feedstocks for energy production (Ou et al., 2022). WCO is another resource generally discarded due to losing its physical and chemical properties. It is a valuable source of biomass because of its presence in abundance globally and its unique properties for fuel production. The researchers meticulously reviewed the possible use of WCO-derived biodiesel, which established it advantageously in the technical, economical, tribological and environmental aspects. It was calculated that the CO_2 equivalent of WCO-derived biodiesel is 85% lower than conventional diesel (Xu et al., 2022).

Herwaarden et al. (2022) studied the potential of a high erucic acid content of Carinata oil (*Brassica carinata*) as a feedstock for renewable diesel production. This crop is mainly grown in Australia and is known for its better heat, drought and pest tolerance. It gives excellent resistance to soil-borne fungal diseases. After extracting oil, the seed meal is heat-treated to capture the remaining solvents and remove volatile contents. Carinata seed meal has low fibre content and high protein, which proves beneficial as animal rations. This by-product has a higher value than the oil itself.

3.2.2 Renewable diesel production pathways

Renewable diesel is composed of a mixture of linear hydrocarbons obtained from the hydrogenation process with a similar chemical composition to fossil fuel. It is derived from biomass and uses a different technique to convert the feedstocks into molecules compatible with fossil diesel. Renewable diesel, also known as green diesel, hydro-processed renewable diesel (HRD) or hydrogenated vegetable oil (HVO), is produced by hydrotreating tri-acylglycerols in feedstocks with hydrogen (Cavalcanti, Ravagnani, Stragevitch, Carvalho, & Pimentel, 2022). Oil refineries use the hydrotreatment to remove contaminants such as sulphur, nitrogen, condensed aromatic rings and metals.

The raw material reacts with hydrogen at high temperatures and pressure to change the chemical composition in the process. In the case of renewable diesel, hydrogen is introduced into the raw material in a catalyst that removes oxygen from the free fatty acids and triglyceride molecules, converting them into paraffinic hydrocarbons. This is made up of three main reactions: hydrodeoxygenation (HDO), decarbonylation (DCN) and decarboxylation (DCX), which remove oxygen, carbon monoxide, water and CO_2, respectively (Bacon, 1955).

Hydrothermal deoxygenation (HTDO) of oxygen-containing feedstock such as fatty acid and lipids produces diesel-like hydrocarbons under mild conditions. The process solves the problem of difficult product separation caused by organic solvents. However, it also requires high temperature and pressure reaction conditions, thus making it uneconomical. The application of carbon-supported noble metal catalysts (Pd/C, Ru/C, etc.) promotes the DCX of higher aliphatic esters into alkanes at higher temperatures and pressure (Cai et al., 2022). One of the promising options for obtaining microbial biomass for renewable diesel production is the hydrothermal liquefaction (HTL) process. The primary advantage of the HTL process is the use of wet biomass. Still, despite following a wet route, higher diesel yields were obtained with less humid biomass, indicating the need for initial biomass dewatering. Compared to other processes, HTL converts the entire organic content in the biomass to diesel instead of just the lipid or carbohydrate fraction. The calorific value of the diesel produced by HTL varies between 22 and 39 MJ/kg (Marangon, Castro, Assemany, Couto, & Calijuri, 2022).

A similar way of producing renewable diesel from vegetable oil is catalytic cracking. Cracking reaction with the vegetable oil containing triglycerides into free fatty acid takes place above 500 K, and complete volatilisation takes place above 600 K. The maximum decomposition takes place at 685 K. Two critical reactions occur after the decomposition due to thermal effects: the scission of C–C bonds and the deoxygenation of fatty acids. The deoxygenation can be DCN, decarboxylation or HDO reaction. The acid site may accelerate the cracking of the C–C bonds in the cracking process (Lee & Wang, 2022; Zandonai et al., 2016) produced an HRD using a $NiMo/\gamma-Al_2O_3$ as a catalyst from a feedstock of crude palm oil. The spent catalyst was regenerated with the supercritical CO_2 and cosolvent. In the DCX reaction, a CO_2 molecule is separated from the carboxyl group in the fatty acid structure. In contrast, in the DCN reaction, the same carboxylic group is separated as CO and water in the form of *n*-alkanes. Oxygen is separated via C=O bond hydrogenation, C–O bond rupture and C–C bond cleavage in the HDO reaction. The high porosity and the existence of rich Lewis acid sites in the supported catalyst boost the HDO reaction's performance and the diesel-range selectivity (Nur Azreena et al., 2022).

Spatari et al. (2020) conducted experiments on two pathways for producing renewable diesel from forest residue feedstocks. One is the catalytic fast pyrolysis (CFP) with hydrocracking, and the second is the fast pyrolysis (FP) with catalytic upgradations via hydrotreating and hydrocracking, termed in situ and ex situ catalytic upgradation, respectively. The upgradation of FP bio-oil to oxygen-free fuels requires a significant amount of hydrogen and catalysts for hydrotreating and hydrocracking because of the degradation of catalysts and their need to regenerate during continuous production. Pragya and Pandey (2016) studied the wet and dry routes, open raceway ponds or photobioreactor routes to produce renewable diesel from microalgae biomass. The renewable diesel production via open raceway pond following the dry and wet route has less negative net energy balance (NEB) and a high net energy ratio (NER) than the photobioreactors. The comparative study suggested that the open raceway pond dry route has a higher NER than the wet route.

3.2.3 Alternative pathways and comparison of synthetic diesel, biodiesel and renewable diesel production

The legislative and regulatory bodies of different countries have addressed the intensified concerns for energy sources such as feedstock availability, price volatility, depletion of natural reserves and GHG emission. Hence, fuel production from various renewable sources is gaining much importance, especially fuels derived from biological sources and lipids materials.

Synthetic diesel, petrodiesel or diesel fuel derived from petroleum is the conventional fuel used in automobiles. Synthetic diesel standards are provided by the American Society for Testing and Materials (ASTM D6584) in the United States and European Committee for Standardization (EN 590) in Europe (Knothe, 2010). Biodiesel is a fuel comprising mono-alkyl esters of long-chain fatty acids derived from vegetable or animal fats. It is designated as B100 in the ASTM D6751 standards and EN 14214 as per the European standards (Taghavifar & Mardani, 2015). This category of fuel is generally obtained from the oil or fat by transesterification reaction, yielding glycerol as a by-product. The carbon atoms in biodiesel are approximately 95% "bio" as the methanol used as the alcohol component in its production is usually derived from non-renewable natural gas. In the case of ethyl esters, it is 100% "bio" as the ethanol used is derived from renewable sources such as corn and sugarcane (Ardolino, Parrillo, & Arena, 2018).

The significant components of biodiesel are the straight-chain fatty acids containing 16 and 18 carbon atoms. These feedstocks also contain shorter chain acids, such as lauric acid, other than the typical C16 and C18 acids (Knothe, 2010). Renewable diesel is comprised of mainly long-chain alkanes having a higher cetane number. The cloud point of renewable diesel in the exhaust emission is 7°C, with the value depending upon the severity of the production process. The renewable diesel with better cold flow properties would exhibit lower cetane numbers as the molecules required for improved cold flow have lower cetane number (Spugnoli, Dainelli, D'Avino, Mazzoncini, & Lazzeri, 2012). Both biodiesel and renewable diesel production require a catalyst and co-reagent to obtain the product from the feedstock. Biodiesel requires alcohol (mostly methanol to give methyl esters) as a co-reagent, and renewable diesel requires hydrogen to saturate double bonds and replace oxygen. The most commonly used catalyst in biodiesel production is sodium alkylate (methylate in the case of methanol) or sodium or potassium hydroxide (Mila, Canals, Burnip, & Cowell, 2006).

The molecular weight of the feedstock plays a vital role in maintaining the molar ratio of 6:1 of alcohol/vegetable oil for biodiesel. The fatty acid profile plays a vital role as the saturation of fatty acid chains influences the amount of hydrogen consumed in the case of renewable diesel (Leung et al., 2010). Highly saturated feedstocks such as palm oil and animal fats are advantageous in reducing hydrogen requirements and reducing production costs. The essential advantage of renewable diesel production is that it requires no dedicated facility for processing the feedstock, as it can be processed in the petrochemical production stream (Pehnt, 2006). There is a significant difference between the energy balance of biodiesel and renewable diesel due to the difference in the production process and the energy content of the products. In biodiesel, the energy balance is altered by the production of alcohol, as each alcohol has its energy requirement. The amount of hydrogen varies the energy balance depending on the desired product in the case of renewable diesel (Sunde, Brekke, & Solberg, 2011).

Conventional diesel utilised in automobiles has a primary challenge of exhaust emissions such as nitrogen oxides (NOx), particulate matter (PM), hydrocarbons and carbon monoxide

(CO). A work conducted by Gmunder et al. (2010) showed that the components such as hexadecane and do decane reduced the NOx emission by 15%–16% more than the biodiesel components. The PM emission was reduced by 73%–83% by biodiesel components rather than alkanes. The hydrocarbon emissions were strongly affected by the chain length of the alkanes, which performed better than the biodiesel. The CO emissions were decreased significantly by biodiesel rather than alkanes. Biodiesel performs better in PM and CO emissions than renewable diesel. Biodiesel and renewable diesel biodegrade quickly as compared to synthetic diesel due to their composition when exposed to suitable conditions and microorganisms. The cultivation and economics of biodiesel and renewable diesel are identical from the point of production as both use the same feedstock. The difference between the two can be attributed to the postproduction distribution (Ally & Pryor, 2007).

3.3 Evaluation of the life cycle assessment process of renewable diesel and product assessment

Renewable diesel has emerged as an alternate and renewable energy source to reduce carbon footprints. The LCA is performed for the growth and evaluation of renewable diesel and to evaluate the energy and economic output of various feedstocks used to produce renewable diesel. Kumar et al. (2022) conducted comparative energy and economic analyses using the LCA approach to produce renewable diesel from various vegetable oils such as Jatropha, Mahua, Neem, Palm, Coconut, Karanja, Jojoba and Tung. The study was conducted according to the ISO standards, comprising crop cultivation, oil extraction and renewable diesel production. The cultivation phase included energy inputs such as agricultural machinery, human labour, fertiliser, chemical, fossil fuel, seeds, electricity and water for irrigation. The oil extraction phase included seed decortication, expelling oil from seeds and oil filtration. The renewable diesel production phase included the esterification and transesterification process that require energy inputs such as alcohol, catalyst, machinery, organic fertiliser, water, electricity and human labour. The energy output included renewable diesel, crude glycerol and the seed/oil cake. The results of all the different feedstocks indicated that the net energy is positive for all the feeds. The neem oil had the highest energy ratio of 5.5164 and energy productivity of 0.0567 MJ/kg. The values obtained for neem were as follows: gross production value (Rs. 222.27/kg), gross returns (Rs. 168.70/kg), net returns (Rs. 144.13/kg), benefit-to-cost ratios (2.84) and productivity (0.0122 kg/Rs.). Thus neem can be a higher contributor to reducing CO_2 emission (g CO_2eq/MJ) in diesel blending.

Pragya and Pandey (2016) produced renewable diesel from the microalgal biomass following the dry and wet route, giving a positive NEB. The photobioreactors produce much more biomass than the open pond in the same period. The extra energy is compensated by the high energy requirements of the tube lights and heat exchangers. The open pond has a higher NER than photobioreactors. The NER of the dry route is higher than the wet route as the CO_2 emission is higher. This may be due to the higher energy requirements during HTL in the wet route. The total energy use is five times more than the energy output with a lower NER value of 0.20 by the best possible route. The value of NEB is -4.07 MJ as it uses more energy to produce per MJ of energy.

Spatari et al. (2020) performed a comparative study between the cost and environmental life cycle performance of the renewable diesel produced from forest residues by the CFP and FP methods. CFP requires upgrading in one catalytic hydrocracking step due to the integration of zeolite catalysts. The FP requires upgrading by multiple catalytic hydrotreating and hydrocracking steps, increasing the capital cost. The selling price of FP and CFP fuel is ($1.69/L) and ($2/L), respectively. Despite having lower capital costs, the higher selling price of CFP fuel is due to the higher operating costs from hydrogen purchase and catalyst replacement. Utilities, hydrogen and catalyst contribute to GHG emissions during both processes. The GHG emission for CFP and FP are (-72 to 32 g CO_2e/MJ) and (8–13 g CO_2e/MJ), respectively. The FP fuel is more favourable due to the higher biofuel yield, low emissions, compliance with energy policy objectives and economics.

Dang, Yu, and Luo (2014) conducted a comparative LCA analysis of renewable diesel production from corn stover by pyrolysis and successive hydrotreating and hydrocracking methods. The study was conducted based on the Greenhouse Gases, Regulated Emissions, and Energy Use in Transportation (GREET) model. Three different cases were explored for renewable diesel upgradation: In Case 1, the hydrogen produced by natural gas is consumed, and all of the bio-oil produced by FP is upgraded to gasoline and diesel. In Case 2, hydrogen for hydroprocessing is obtained by steam reforming 35% of the aqueous phase, while the remaining bio-oil is hydrogenated to produce gasoline and diesel. In Case 3, hydrogen is produced by steam reforming all 62.1% of the aqueous phase of bio-oil, while pyrolytic lignin is hydro-processed for liquid biofuel production. The environmental impacts, such as non-renewable energy demand (NRED) and global warming potential (GWP), were evaluated for 1 MJ of biofuel produced. The percentage reduction in the NRED in all three cases is 67.5%, 76.7% and 147.5%, respectively. The reduction in GWP in all three cases is 69.1%, 73.0% and 119.4%, respectively, making the third case the most efficient in decreasing the environmental impacts. The biofuel yields of the three cases were 313.2%, 23.7% and 11.5%, respectively, making the first case the most efficient in renewable diesel production.

Talens Peiro et al. (2010) and Talens Peiro, Lombardi, Villalba Méndez, Gabarrell, and Durany (2010) conducted an LCA and energetic life cycle assessment (ELCA) of renewable diesel production from UCO. The life cycle consisted of four stages: (1) collection, (2) pretreatment, (3) delivery and (4) transesterification of UCO. The 1-tonne renewable diesel production was assessed for environmental impact by LCA and the ELCA to account for the exergy input of the system. The transesterification stage had more impact, between 63% and 90%, than other stages in 8 out of 10 impact categories (abiotic depletion, GWP, ozone layer depletion, human toxicity, freshwater, marine aquatic ecotoxicity, terrestrial ecotoxicity, photochemical oxidation, acidification and eutrophication). Prior to converting the energy and mass into exergy units, the inputs were grouped into nine groups for the ELCA (coal, natural gas, uranium, wood, crude oil, wind, solar, biomass and barrage energy). The final results obtained from the study showed that the transesterification stage caused 68% of the total environmental impact, and the main exergy inputs were natural gas and uranium.

Fan, Shonnard, Kalnes, Johnsen, and Rao (2013) studied the efficiency of field pennycress (*Thlaspi arvense* L.), a member of the mustard family, in producing biomass for renewable diesel. They investigated the LCA for GHG emissions and energy balance for pennycress-derived hydro-processed renewable jet (HRJ) fuel and renewable diesel. The life cycle GHG emissions depending upon the coproduct for HRJ and renewable diesel, are (−18 to 45 g CO_2e/MJ) and (13–41 g CO_2e/MJ), respectively. The cumulative energy demand (CED) was divided into four parts: non-renewable fossil and nuclear energy use, renewable biomass energy use, and other renewable energy sources such as solar, geothermal and hydroelectric power. The CED varied from 1.3 to 1.9 and 1.3–2.3 MJ/unit for renewable diesel and HRJ. The fossil energy demand (FED) was divided into seven stages: cultivation, raw material acquisition, feedstock transportation, oil extraction, fuel production, distribution and use. HRJ and renewable diesel FED ranged from −0.7 to 0.4 and −0.1 to 0.4 MJ/unit. Compared with petroleum and syndiesel, the energy demand for renewable diesel and HRJ was significantly lower.

Okeke et al. (2020) and Okeke, Sahoo, Kaliyan, and Mani (2020) conducted the LCA study to produce drop-in renewable diesel derived from miscanthus from biogas via solid-state anaerobic digestion. The analysis included five stages: miscanthus cultivation, harvest, transportation to the biorefinery, conversion of biomass to Fischer–Tropsch diesel and combustion of the renewable in a passenger vehicle. The result showed a 73% decrease in GHG emissions with miscanthus-derived renewable diesel than fossil-derived diesel. A 110% reduction in GHG emissions was observed when the biomass yield dropped from 10 to 5 Mg/ha. The acidification potential for the renewable diesel was obtained as (0.22 kg SO_2eq/GGE), which was 91% higher when compared to the conventional diesel (0.02 kg SO_2eq/GGE). The fossil fuel depletion of the renewable diesel process was 4.91 MJ/GGE, which is 74% lower than that of conventional diesel. The total energy efficiency of renewable diesel was calculated to be 33.99%, which is at par with other fuels, proving the process's feasibility.

3.4 Select case studies on life cycle assessment of renewable diesel production

A few specific LCA and LCI (Life Cycle Impact) assessment studies for renewable diesel production from various feedstocks are discussed below, focusing mainly on the steps included in the study, the methodology followed, the study's purpose, and the outcome.

3.4.1 Life cycle assessment of renewable diesel production from lignocellulosic biomass (Wong, Zhang, & Kumar, 2016)

The study discusses the LCA of hydrogenation-derived diesel (HDRD) production from the lignocellulosic biomass in Canada.

The objective includes the development of the model to estimate the energy input to produce HDRD from the lignocellulosic biomass. The first step of LCA is well to wheel approach to analyse whether the HDRD process is an environmentally friendly process for the reduction of GHGs. The LCA study comprised NER to determine the ratio of energy output to fossil fuel energy input. The scope of the work includes the emissions and energy usage from various stages of the HDRD process. The functional unit comprised 1 MJ of the renewable diesel produced based on the lower heating value of the fuel. The system boundary is comprised of direct inputs of fossil fuel in every stage of HDRD production for the complete LCA process.

The LCI step is comprised of how the biomass is obtained and comprised of feedstocks such as wheat, barley, oats, etc., and the next step is biomass harvesting and collection. The step comprised of transportation of tree chips and forest residue. The step of FP is comprised of the thermal decomposition of biomass in the absence of air and is used to produce the bio-oil. The study assumes that the distances from the pyrolysis plants to the HDRD plant are estimated to be 300 km for whole trees and forest residue and 250 km for the agricultural residue. The results include the allocation methods, assumptions and system boundaries. The study showed that harvesting feedstock from agricultural residue is more efficient than harvesting feedstock. The higher NER value of agricultural residue is NER of 1.90 MJ/MJ as compared to the NER values of 1.71 and 1.55 MJ/MJ, respectively. This relates to the lower GHGs (Table 3.1).

The feedstock determines the control values for the feasibility of producing HDRD. The study showed that if the forest lands are not fertilised after the harvesting, GHG emissions of the whole feedstock will be 37.6 g CO_2 eq./MJ HDRD with a NER of 1.77 MJ/MJ if the nutrient return is not carried out. The final result of the complete analysis shows that the GHG emissions can be 39%–47% of those of petroleum diesel alternatives. This indicated that renewable diesel could mitigate GHG emissions. The major energy consumer in the HDRD process is the step of hydroprocessing. The agricultural feedstock has a higher harvesting efficiency than the whole tree harvesting process to lower energy usage and GHG emissions (Tables 3.2 and 3.3).

Table 3.1 LCA of production of renewable diesel from sunflower biomass (Spinelli, Jez, Pogni, & Basosi, 2013).

Step	Methodology	Purpose	Findings-highlights
Goal		Chain of pilot production of biodiesel to reduce possible local fossil fuel consumption	
Scope definition	Includes the agricultural phase, sunflower oil extraction, transesterification and all transportation within the system boundary	Production of renewable diesel from sunflower	
Functional unit	1 kg of renewable diesel from sunflower oil		
Material flow accounting (MFA)	The material intensity factors (g/unit) are multiplied by each input, respectively, which accounts for the total abiotic matter, water and air required to provide the information to the system	Accounts for the material flow directly and indirectly involved in the process and evaluates the environmental disturbances associated with the withdrawal of resources from their natural ecosystem	The total amount of abiotic matter per kg of biodiesel was 5.64E + 03 g/kg The total amount of water per kg of biodiesel was 6.52E + 03 g/kg The total amount of air matter per kg of biodiesel was 2.50E + 02 g/kg Cumulative material consumption was 1.25E + 04 g/kg
Embodied energy analysis	The system's energy requirements are calculated as the sum of the individual oil equivalents of the input flows. It can be converted to an energy unit by multiplying the standard calorific value of diesel oil	All material and energy flows supporting each step of the system are accounted for according to the specific cumulative energy cost	Production of 1 kg of biodiesel required 2.22E + 02 g of crude oil as energy input The CO_2 emission for 0.71 kg of biodiesel was 3.18 kg CO_2/kg oil fuel according to combustion stoichiometry
Energy accounting	It uses solar energy embodied in the system's input as a measurement base. The total amount of solar energy available is directly or indirectly required to make a product or support a process.	Evaluates the environmental performance of a system on a global scale	Specific energy for sunflower was 3.29E + 12 seJ/kg Specific energy for biodiesel production in the industrial phase was 7.24E + 12 seJ/kg Converting specific emergy to transformity, the resulting value was 1.00E + 05 seJ/kg The environmental loading ratio indicated that the non-renewable energy was 1.22 times higher than the renewable energy for biodiesel production The renewability of renewable diesel was 45%
LCA (eco-indicator 99 method)	Includes the analysis through a damage evaluation step and estimates various environmental impacts such as damage to human health expressed as DALY (Disability Adjusted Life Years), damages to the ecosystem and resource extraction related to the quality of remaining mineral and fossil resources	Provides the life cycle impact assessment and life cycle interpretation. This method is an LCA weighting method developed for product design and is easily understandable and user friendly	The estimate of uncertainty was highest for carcinogens 2.74E-09 DALY. Other environmental impact categories had uncertainty in the range of 5%–17% Damage to human health, ecosystem quality and resources expressed for biodiesel as points (Pt) were 7.09E-03, 2.99E-01 and 7.26E-02, respectively The result established that 1 kg of biodiesel releases 0.71 kg of CO_2, which is 1.94E + 01 g CO_2/MJ

Table 3.2 LCA of renewable diesel production from palm oil (Yee et al., 2009).

Step	Methodology	Purpose	Findings-highlights
Goal		Assessment of the energy balance and GHG emissions related to the production of renewable diesel from palm oil	
Scope definition	Includes starting from the plantation of oil palm (*Elaeis guineensis*) stage to the combustion of the renewable in diesel engines	Production of renewable diesel from palm oil in Malaysia	
Functional unit	Unit for energy is GigaJoule (GJ)/tonne palm oil (tCPO)/year Unit for GHG evaluation is tonne CO_2/tonne biodiesel/year	Calculations were done by utilising the energy coefficients reported in various literature	
System boundary	It is divided into three stages first stage—agricultural/plantation phase second stage—palm oil milling stage third stage—transesterification process of palm biodiesel production	Feasibility and Environmental Impact Assessment (EIA) studies were required for developing new plantations above 500 ha on a primary forest or plantation Fresh fruit bunches (FFB) harvested from oil palm trees via sterilisation, threshing and stripping of fruits, digestion and oil extraction Transesterification reaction with methanol in batch-type reactor using alkaline sodium hydroxide	Harvest and collection of FFB can be done up to 24–30 years after planting, depending upon the soil type and practices engaged
Energy balance assessment	In the agricultural phase, the use of fossil fuels in traction and transportation of fertilisers and pesticides was involved A substantial amount of electricity and steam was needed to obtain the required CPO (crude palm oil) in the mill stage. For heat or power production, biomass was used through direct combustion	Used as a techno-economic and environmental feasibility analysis index, the energy produced (output/tonne biodiesel) to the energy consumed (input/tonne biodiesel) per unit of product in palm biodiesel production. It starts from the oil palm plantation till the combustion of palm diesel in diesel engines of vehicles	The total annual energy utilised for fertilisers was 1.76 GJ/tonne CPO The total energy content of fibre and shell was approximately 3.9 GJ/tonne CPO annually Annual electricity of 0.27 GJ/tonne of CPO was utilised in the process of producing CPO The total energy used in the transesterification process was 1.80 GJ/tonne CPO annually The total energy released due to the combustion of palm diesel was 39.2 GJ/tonne CPO annually The total energy contained in the biodiesel, glycerol and biomass was 60.72 GJ/tonne CPO annually

(*Continued*)

Table 3.2 (Continued)

Step	Methodology	Purpose	Findings-highlights
GHG assessment	In the plantation phase, it was assessed that oil palm crops emit 8–10 times more oxygen and absorb up to 10 times more CO_2/hectare/year The use of fertilisers and traction activities also release CO_2 in the milling phase In the transesterification phase, the emissions and electricity generation of the steam boilers contribute to significant CO_2 emission	The primary motive of renewable diesel is to replace the use of conventional fuel and reduce GHG emissions arising from that. GHG assessment is to determine the effect of biodiesel use on GHG emissions.	The average CO_2 emission for this study was 42.5 tonnes of CO_2/tonne of CPO 211,996.97 kg CO_2 released per tonne of biodiesel produced The utilisation of fertiliser and traction activities released 11,630.88 kg CO_2 and 5872.81 kg CO_2/tonne of biodiesel The incineration of fibre and shell produced 117,234.33 kg CO_2/tonne of biodiesel In the transesterification process, 2087.26 kg CO_2/tonne of biodiesel was released
LCI	Provides the input and output of all the stages involved in the renewable palm biodiesel production The products and co-products produced by the palm oil mill The conversion of CPO to biodiesel, annual production, energy coefficient of fuel utilised and calorific value of products	The inventory comprises all the recognised inputs and outputs to and from the system boundary	The agricultural stage produced 19 tonnes of FFB for each hectare of plantation land It yielded 0.20 tonnes of CPO for each tonne of FFB processed The yield of co-products produced (fibre and shell, empty fruit bunches and palm oil mill effluent) were 190 kg, 230 kg and 600–700 kg/tonne of FFB, respectivelyThe efficiency of conversion of CPO to biodiesel was 99% The annual production was 126,888 tonnes or 0.99 tonnes of biodiesel per tonne of CPO feedstock

Table 3.3 LCA of renewable diesel production from miscanthus via anaerobic digestion (Okeke et al., 2020).

Step	Methodology	Purpose	Findings-highlights
Goal		Systematic assessment and evaluation of the environmental and energy impact of the production of drop-in renewable diesel from biogas produced by anaerobic digestion of miscanthus cultivated from the strip-mined lands	
Scope definition	It included emissions from the energy input, harvest and logistics, anaerobic	Production of renewable diesel from biogas obtained from the anaerobic	

(Continued)

Table 3.3 (Continued)

Step	Methodology	Purpose	Findings-highlights
	digestion of pretreated miscanthus to biogas, digestate utilisation, transportation, biogas cleaning, biomethane conversion to syngas, catalytic conversion of syngas to Fischer–Tropsch (FT) syn-crude, electricity generation, FT syn-crude upgrade to renewable diesel and drop-in diesel combustion	digestion of miscanthus from strip-mined lands at Ohia, United States	
Functional unit	The analysis was done in one gasoline gallon equivalent (GGE) of drop-in diesel	Calculations were done comparing it against the conventional petroleum-derived diesel	
System boundary	The system consists of the cultivation of miscanthus for aerobic digestion for the production of biogas being converted to drop-in renewable diesel to be used in a vehicle	The process followed the process from feedstock cultivation to miscanthus harvesting and logistics, feedstock pretreatment, anaerobic digestion, and biogas to drop-in diesel combustion. All the thermal and electrical energy used in the system was generated internally, avoiding the use of any external energy supply	The energy allocation method was employed for emission distribution among the liquid fuels. The % emission among liquid fuels were 6%, 32% and 62% for LPG, gasoline and drop-in diesel, respectively. In terms of energy produced, 5.7%, 5.6%, 30.2% and 58.5% were contributed by electricity, LPG, gasoline and drop-in diesel, respectively, from the total output energy of the system
LCI	LCI was conducted based on ISO 14000 environmental standards. SimaPro 8, a life cycle analysis tool, and TRACI 2.1 v4, an impact calculation tool, were utilised to estimate the emissions accredited to the life cycle of the drop-in renewable diesel	All inputs (raw materials and energies) required to produce the drop-in renewable diesel with the by-products and emissions (air, water and land) were accounted for. The Aspen Plus model and GREET model were used for emissions, mass and energy balance	0.032 dry mg of miscanthus was digested to produce 7.7 kg (6.47 Nm^3) of biogas, equivalent to 1 GGE of drop-in diesel
LCIA	The environmental impacts were assessed by TRACI 2.1 v4 by US EPA to produce drop-in renewable diesel from miscanthus	The assessment of the natural resources used, water, energy and their environmental impacts	The impacts were ranked based on global, regional and local impact and their magnitude of impact The acidification potential drop-in diesel was 0.22 kg SO_2eq/GGE The drop-in diesel had 5% and 10% carcinogenic (2.75×10^{-8} CTUh) and non-carcinogenic (2.64×10^{-7} CTUh) tendencies, respectively Respiratory effects of drop-in diesel were 7.44×10^{-3} kg $PM_{2.5}$/GGE The fossil fuel depletion of drop-in diesel was 4.91 MJ/GGE The solar radiation efficiency of drop-in diesel is 1.71%

(*Continued*)

Table 3.3 (Continued)

Step	Methodology	Purpose	Findings-highlights
Sensitivity Analysis	The analysis was limited only to the base case design by varying the system boundary parameters, which are sensitive to the entire process	It determined the effect of variations in the critical input parameter values on the environmental impacts of the proposed system	The increase and decrease of 50% in the nitrogen input during miscanthus cultivation caused a 12% increase and 13% decrease in GHG emissions, respectively The methane yield of 100 LCH_4/kg_{VS} during anaerobic digestion caused a reduction of 64% in GHG The drop in biomass yield from 10 Mg/ha to 5 Mg/ha caused a 110% reduction in GHG
Scenarios Analysis	Three scenarios analysed were Miscanthus cultivation with lime application Miscanthus cultivation without lime application Digestate reuse to displacement chemical fertiliser during miscanthus cultivation	Different scenarios were considered for overall analysis to get a better understanding of the environmental performances during the production of renewable diesel	The second and third scenarios decreased the GWP to 22% and 24%, respectively, compared to the first scenario The third scenario showed a 7% and 10% reduction in carcinogenic and non-carcinogenic tendencies, respectively The third scenario reduced the ecotoxicity and fossil fuel depletion impacts by 11% and 22%, respectively, compared to the first scenario No substantial reduction in eutrophication potential, acidification and ozone depletion

3.5 Conclusion

The LCA studies discuss renewable diesel production from lignocellulosic biomass, sunflower oil and palm oil. The HDRD study from the lignocellulosic biomass concludes that the GHG emissions can be 39%–47% of these petroleum diesel alternatives. The study shows that 3.18 kg of CO_2/kg of oil fuel for 0.71 kg of biodiesel. The efficiency of the conversion of palm oil to biodiesel production was obtained as 99%. The work on renewable diesel can be further studied, and the LCI study can be included to check the feasibility of renewable diesel scaled-up production.

References

Aatola, H., Larmi, M., Sarjovaara, T., & Mikkonen, S. (2009). Hydrotreated vegetable oil (HVO) as a renewable diesel fuel: Trade-off between NO_x, particulate emission, and fuel consumption of a heavy duty engine. *SAE International Journal of Engines, 1*(1), 1251–1262. Available from https://doi.org/10.4271/2008-01-2500.

Ally, J., & Pryor, T. (2007). Life-cycle assessment of diesel, natural gas and hydrogen fuel cell bus transportation systems. *Journal of Power Sources, 170* (2), 401–411. Available from https://doi.org/10.1016/j.jpowsour.2007.04.036.

Ardolino, F., Parrillo, F., & Arena, U. (2018). Biowaste-to-biomethane or biowaste-to-energy? An LCA study on anaerobic digestion of organic waste. *Journal of Cleaner Production, 174*, 462–476. Available from https://doi.org/10.1016/j.jclepro.2017.10.320.

Bacon, H. P. (1955). Animal, vegetable or mineral? *Journal of Geography, 54*(2), 97–98. Available from https://doi.org/10.1080/00221345508982847.

Bezergianni, S., & Dimitriadis, A. (2013). Comparison between different types of renewable diesel. *Renewable and Sustainable Energy Reviews, 21*, 110–116. Available from https://doi.org/10.1016/j.rser.2012.12.042.

Canter, C. E., Davis, R., Urgun-Demirtas, M., & Frank, E. D. (2014). Infrastructure associated emissions for renewable diesel production from microalgae. *Algal Research, 5*(1), 195–203. Available from https://doi.org/10.1016/j.algal.2014.01.001.

Cavalcanti, C. J. S., Ravagnani, M. A. S. S., Stragevitch, L., Carvalho, F. R., & Pimentel, M. F. (2022). Simulation of the soybean oil hydrotreating process for green diesel production. *Cleaner Chemical Engineering, 1*(February), 100004. Available from https://doi.org/10.1016/j.clce.2022.100004.

Centi, G., & Van Santen, R. A. (2008). Catalysis for renewables. From feedstock to energy production. Edited by Gabriele Centi and Rutger A. van Santen. *ChemSusChem, 1*(3), 262. Available from 10.1002/cssc.200700153.

Chakraborty, R., Gupta, A. K., & Chowdhury, R. (2014). Conversion of slaughter house and poultry farm animal fats and wastes to biodiesel: Parametric sensitivity and fuel quality assessment. *Renewable and Sustainable Energy Reviews, 29*, 120–134. Available from https://doi.org/10.1016/j.rser.2013.08.082.

Chen, R., Qin, Z., Han, J., Wang, M., Taheripour, F., Tyner, W., ... Duffield, J. (2018). Life cycle energy and greenhouse gas emission effects of biodiesel in the United States with induced land use change impacts. *Bioresource Technology, 251*, 249–258. Available from https://doi.org/10.1016/j.biortech.2017.12.031.

Chia, S. R., Nomanbhay, S., Ong, M. Y., Shamsuddin, A. H., Bin., Chew, K. W., & Show, P. L. (2022). Renewable diesel as fossil fuel substitution in Malaysia: A review. *Fuel, 314*(December 2021), 123137. Available from https://doi.org/10.1016/j.fuel.2022.123137.

Cai, X., Zhang, Z., Ye, Y., Wang, D., Li, S., Wang, D., & Zheng, Z. (2022). Conversion of higher fatty acids or higher fatty acid esters to long-chain alkanes by acid added metal catalyst under mild hydrothermal conditions. *Biomass and Bioenergy, 156*(December 2021), 106328. Available from https://doi.org/10.1016/j.biombioe.2021.106328.

Dang, Q., Yu, C., & Luo, Z. (2014). Environmental life cycle assessment of bio-fuel production via fast pyrolysis of corn stover and hydroprocessing. *Fuel, 131*(2014), 36–42. Available from https://doi.org/10.1016/j.fuel.2014.04.029.

Demirbas, A. (2009). Progress and recent trends in biodiesel fuels. *Energy Conversion and Management, 50*(1), 14–34. Available from https://doi.org/10.1016/j.enconman.2008.09.001.

Fan, J., Shonnard, D. R., Kalnes, T. N., Johnsen, P. B., & Rao, S. (2013). A life cycle assessment of pennycress (*Thlaspi arvense* L.)-derived jet fuel and diesel. *Biomass and Bioenergy, 55*, 87–100. Available from https://doi.org/10.1016/j.biombioe.2012.12.040.

Guzman, A., Torres, J. E., Prada, L. P., & Nuñez, M. L. (2010). Hydroprocessing of crude palm oil at pilot plant scale. *Catalysis Today, 156*(1–2), 38–43. Available from https://doi.org/10.1016/j.cattod.2009.11.015.

Gmunder, S. M., Zah, R., Bhatacharjee, S., Classen, M., Mukherjee, P., & Widmer, R. (2010). Life cycle assessment of village electrification based on straight jatropha oil in Chhattisgarh, India. *Biomass and Bioenergy, 34*(3), 347–355. Available from https://doi.org/10.1016/j.biombioe.2009.11.006.

Herwaarden, A.Van, Bennett, R., Potter, T., Lambrides, C., Gororo, N., Krakowski, H., & Arief, V. (2022). The performance and feasibility of Carinata in Australia. Proceedings of the 20th Australian Agronomy Conference, Toowoomba, QLD, Australia, September 18–22, pp. 3–6.

Huber, G. W., O'Connor, P., & Corma, A. (2007). Processing biomass in conventional oil refineries: Production of high quality diesel by hydrotreating vegetable oils in heavy vacuum oil mixtures. *Applied Catalysis A: General, 329*, 120–129. Available from https://doi.org/10.1016/j.apcata.2007.07.002.

IEA. (2018). Renewables 2018: Analysis and forecasts to 2023.

Kaewmeesri, R., Srifa, A., Itthibenchapong, V., & Faungnawakij, K. (2015). Deoxygenation of waste chicken fats to green diesel over Ni/Al_2O_3: Effect of water and free fatty acid content. *Energy and Fuels, 29*(2), 833–840. Available from https://doi.org/10.1021/ef5023362.

Kalnes, T. N., Marker, T., Shonnard, D. R., & Koers, K. P. (2008). Green diesel production by hydrorefining renewable feedstocks. *Biofuels Technology, 4*, 7–11.

Knothe, G. (2010). Biodiesel and renewable diesel: A comparison. *Progress in Energy and Combustion Science, 36*(3), 364–373. Available from https://doi.org/10.1016/j.pecs.2009.11.004.

Krár, M., Kovács, S., Kalló, D., & Hancsók, J. (2010). Fuel purpose hydrotreating of sunflower oil on CoMo/Al_2O_3 catalyst. *Bioresource Technology, 101*(23), 9287–9293. Available from https://doi.org/10.1016/j.biortech.2010.06.107.

Kumar, N., Koul, R., & Singh, R. C. (2022). Comparative analysis of ternary blends of renewable diesel, diesel and ethanol with diesel. *Sustainable Energy Technologies and Assessments, 50*(December 2021), 101828. Available from https://doi.org/10.1016/j.seta.2021.101828.

Lapuerta, M., Armas, O., Hernández, J. J., & Tsolakis, A. (2010). Potential for reducing emissions in a diesel engine by fuelling with conventional biodiesel and Fischer–Tropsch diesel. *Fuel, 89*(10), 3106–3113. Available from https://doi.org/10.1016/j.fuel.2010.05.013.

Landau, M. V., Herskowitz, M., Givoni, D., Laichter, S., & Yitzhaki, D. (1998). Medium severity hydrotreating and hydrocracking of Israeli shale oil—II. Testing of novel catalyst systems in a trickle bed reactor. *Fuel, 77*(1–2), 3–13. Available from https://doi.org/10.1016/S0016-2361(97)00173-7.

Lee, C. P., & Wang, W. C. (2022). The production of hydroprocessed renewable diesel with a catalyst regenerated using supercritical carbon dioxide ($SCCO_2$). *Energy, 246*, 123402. Available from https://doi.org/10.1016/j.energy.2022.123402.

Leung, D. Y. C., Wu, X., & Leung, M. K. H. (2010). A review on biodiesel production using catalyzed transesterification. *Applied Energy, 87*(4), 1083–1095. Available from https://doi.org/10.1016/j.apenergy.2009.10.006.

Marangon, B. B., Castro, J. S., Assemany, P. P., Couto, E. A., & Calijuri, M. L. (2022). Environmental performance of microalgae hydrothermal liquefaction: Life cycle assessment and improvement insights for a sustainable renewable diesel. *Renewable and Sustainable Energy Reviews, 155*(May 2021), 111910. Available from https://doi.org/10.1016/j.rser.2021.111910.

Mila, I., Canals, L., Burnip, G. M., & Cowell, S. J. (2006). Evaluation of the environmental impacts of apple production using life cycle assessment

(LCA): Case study in New Zealand. *Agriculture, Ecosystems and Environment, 114*(2–4), 226–238. Available from https://doi.org/10.1016/j.agee.2005.10.023.

McManus, M. C., Taylor, C. M., Mohr, A., Whittaker, C., Scown, C. D., Borrion, A. L., ... Yin, Y. (2015). Challenge clusters facing LCA in environmental decision-making—What we can learn from biofuels. *International Journal of Life Cycle Assessment, 20*(10), 1399–1414. Available from https://doi.org/10.1007/s11367-015-0930-7.

Naik, S. N., Goud, V. V., Rout, P. K., & Dalai, A. K. (2010). Production of first and second generation biofuels: A comprehensive review. *Renewable and Sustainable Energy Reviews, 14*(2), 578–597. Available from https://doi.org/10.1016/j.rser.2009.10.003.

Neste Corporation. (2015). *Neste renewable diesel handbook*, 1–33.

Nur Azreena, I., Lau, H. L. N., Asikin-Mijan, N., Hassan, M. A., Izham, S. M., Safa Gamal, M., ... Taufiq-Yap, Y. H. (2022). Hydrodeoxygenation of fatty acid over La-modified HZSM5 for premium quality renewable diesel production. *Journal of Analytical and Applied Pyrolysis, 161*(December 2021), 105406. Available from https://doi.org/10.1016/j.jaap.2021.105406.

Okeke, I. J., Sahoo, K., Kaliyan, N., & Mani, S. (2020). Life cycle assessment of renewable diesel production via anaerobic digestion and Fischer–Tropsch synthesis from miscanthus grown in strip-mined soils. *Journal of Cleaner Production, 249*, 119358. Available from https://doi.org/10.1016/j.jclepro.2019.119358.

Ou, L., Li, S., Tao, L., Phillips, S., Hawkins, T., Singh, A., ... Cai, H. (2022). Techno-economic analysis and life-cycle analysis of renewable diesel fuels produced with waste feedstocks. *ACS Sustainable Chemistry and Engineering, 10*(1), 382–393. Available from https://doi.org/10.1021/acssuschemeng.1c06561.

Pehnt, M. (2006). Dynamic life cycle assessment (LCA) of renewable energy technologies. *Renewable Energy, 31*(1), 55–71. Available from https://doi.org/10.1016/j.renene.2005.03.002.

Pragya, N., & Pandey, K. K. (2016). Life cycle assessment of green diesel production from microalgae. *Renewable Energy, 86*, 623–632. Available from https://doi.org/10.1016/j.renene.2015.08.064.

Salvachua, D., Saboe, P. O., Nelson, R. S., Singer, C., McNamara, I., del Cerro, C., ... Linger, J. G. (2021). Process intensification for the biological production of the fuel precursor butyric acid from biomass. *Cell Reports Physical Science, 2*(10). Available from https://doi.org/10.1016/j.xcrp.2021.100587.

Simacek, P., Kubička, D., Šebor, G., & Pospíšil, M. (2010). Fuel properties of hydroprocessed rapeseed oil. *Fuel, 89*(3), 611–615. Available from https://doi.org/10.1016/j.fuel.2009.09.017.

Soam, S., & Hillman, K. (2019). Factors influencing the environmental sustainability and growth of hydrotreated vegetable oil (HVO) in Sweden. *Bioresource Technology Reports, 7*, 100244. Available from https://doi.org/10.1016/j.biteb.2019.100244.

Spatari, S., Larnaudie, V., Mannoh, I., Wheeler, M. C., Macken, N. A., Mullen, C. A., & Boateng, A. A. (2020). Environmental, exergetic and economic tradeoffs of catalytic- and fast pyrolysis-to-renewable diesel. *Renewable Energy, 162*, 371–380. Available from https://doi.org/10.1016/j.renene.2020.08.042.

Spinelli, D., Jez, S., Pogni, R., & Basosi, R. (2013). Environmental and life cycle analysis of a biodiesel production line from sunflower in the Province of Siena (Italy). *Energy Policy, 59*, 492–506. Available from https://doi.org/10.1016/j.enpol.2013.04.009.

Spugnoli, P., Dainelli, R., D'Avino, L., Mazzoncini, M., & Lazzeri, L. (2012). Sustainability of sunflower cultivation for biodiesel production in Tuscany within the EU Renewable Energy Directive. *Biosystems Engineering, 112*(1), 49–55. Available from https://doi.org/10.1016/j.biosystemseng.2012.02.004.

Sunde, K., Brekke, A., & Solberg, B. (2011). Environmental impacts and costs of woody biomass-to-liquid (BTL) production and use—A review. *Forest Policy and Economics, 13*(8), 591–602. Available from https://doi.org/10.1016/j.forpol.2011.05.008.

Taghavifar, H., & Mardani, A. (2015). Energy consumption analysis of wheat production in West Azarbayjan utilizing life cycle assessment (LCA). *Renewable Energy, 74*, 208–213. Available from https://doi.org/10.1016/j.renene.2014.08.026.

Talens Peiro, L., Lombardi, L., Villalba Méndez, G., Gabarrell, I., & Durany, X. (2010). Life cycle assessment (LCA) and exergetic life cycle assessment (ELCA) of the production of biodiesel from used cooking oil (UCO). *Energy, 35*(2), 889–893. Available from https://doi.org/10.1016/j.energy.2009.07.013.

U.S. Energy Information Administration. (2019). International Energy Outlook. *Combustion and Flame, 209*, 239–255.

Vibhanshu, V., Kumar, N., Mishra, C., Sinha, S., Pali, H. S., & Bansal, S. (2013). Experimental Investigation of diesel engine fueled with jatropha oil blend with ethanol. *SAE International Journal of Fuels and Lubricants, 6*(3), 951–958. Available from https://doi.org/10.4271/2013-24-0105.

Wang, M., Huo, H., & Arora, S. (2011). Methods of dealing with co-products of biofuels in life-cycle analysis and consequent results within the U.S. context. *Energy Policy, 39*(10), 5726–5736. Available from https://doi.org/10.1016/j.enpol.2010.03.052.

Wong, A., Zhang, H., & Kumar, A. (2016). Life cycle assessment of renewable diesel production from lignocellulosic biomass. *The International Journal of Life Cycle Assessment*, 1404–1424. Available from https://doi.org/10.1007/s11367-016-1107-8.

Xu, H., Li, Y., Li, Z., Song, Y., Zhang, Y., & Song, H. (2022). Methane-assisted waste cooking oil conversion for renewable fuel production. *Fuel, 311*(August 2021), 122613. Available from https://doi.org/10.1016/j.fuel.2021.122613.

Yee, K. F., Tan, K. T., Abdullah, A. Z., & Lee, K. T. (2009). Life cycle assessment of palm biodiesel: Revealing facts and benefits for sustainability. *Applied Energy, 86*(Suppl. 1), S189–S196. Available from https://doi.org/10.1016/j.apenergy.2009.04.014.

Zandonai, C. H., Yassue-Cordeiro, P. H., Castellã-Pergher, S. B., Scaliante, M. H. N. O., & Fernandes-Machado, N. R. C. (2016). Production of petroleum-like synthetic fuel by hydrocracking of crude soybean oil over ZSM5 zeolite—Improvement of catalyst lifetime by ion exchange. *Fuel, 172*, 228–237. Available from https://doi.org/10.1016/j.fuel.2015.12.059.

Zhang, B., & Seddon, D. (2018). *Hydroprocessing catalysts and processes: The challenges for biofuels production* (pp. 1–305). World Scientific.

Zhang, B., Wu, J., Yang, C., Qiu, Q., Yan, Q., Li, R., ... Wu, J. (2018). Recent developments in commercial processes for refining bio-feedstocks to renewable diesel. *BioEnergy Research, 11*, 689–702.

4

Challenges and opportunities in the commercialisation of renewable diesel

Adewale Adewuyi
Department of Chemical Sciences, Faculty of Natural Sciences, Redeemer's University, Ede, Osun State, Nigeria

4.1 Introduction

Renewable diesel (RD) is a biologically sourced fuel that can be used in diesel engines with the capacity to meet specifications such as American Society for Testing and Materials (ASTM) D975 and EN 590 for petroleum in the United States and Europe, respectively. RD has properties which are similar to those of petroleum diesel. RD is available in different parts of the world with the establishment of different production plants that can produce millions of gallons per year. However, this production is expected to grow in the near future. Recently, the United State Environmental Protection Agency (EPA) reported data which indicated the consumption of over 960 million gallons in 2020 (Alternative Fuels Data Center, 2020). RD is also known as green diesel obtained from pyrolysis, hydrotreating, gasification and other biochemical processes (i Nogué et al., 2018; Okeke, Sahoo, Kaliyan, & Mani, 2020), which makes it different from biodiesel (mono-alkyl ester of fatty acids).

In some cases, hydroprocessing may be performed on triglyceride to produce RD, which may involve decarbonylation, hydrodeoxygenation and decarboxylation (Veriansyah et al., 2012). RD is produced from feedstocks such as plants seeds, microalgae and lignocellulosic biomass (Bharath et al., 2021; Low et al., 2021; Oladipupo Kareem et al., 2021). Recent approaches have used novel catalytic processes to enhance feedstock conversion to RD (Ameen, Azizan, Yusup, & Ramli, 2017; Chia et al., 2021), promoting products with better oxidation stability, lower oxidation stability, lower NO_x emissions and good engine performance. Pd and Pt are common examples of catalysts used due to their high

Renewable Diesel. DOI: https://doi.org/10.1016/B978-0-323-91153-5.00007-8

selectivity, but their high cost and scarcity have limited their use. Moreover, the rapid deactivation of Pd limits its use in the production of RD (Cheah et al., 2020). Therefore, the effort focused researchers on finding more suitable and sustainable catalysts for the production of RD.

Several materials have shown potential as viable sources for RD production; among these materials, agricultural wastes and forest residues contain lignocellulose as the main contributory component for RD production. A study revealed that levulinic acid is obtainable from the hydrolysis–dehydration of cellulose with a conversion capacity of approximately 61%; furthermore, about 56% of furfural was also reported from the same process (Han, Sen, Alonso, Dumesic, & Maravelias, 2014). Selective hydrogenation of furfural gives rise to 2-methylfuran (2-MF). Efforts have been documented in the industrial production of 2-MF by hydrolysis–dehydration of hemicellulose sourced from forest residues and agricultural wastes (Nakagawa, Tamura, & Tomishige, 2013). However, several methods have been developed for the alkylation of 2-MF and carbonyl molecules (such as alcohols, alkanal, alkenone, etc.) derived from lignocellulose; such alkylation reaction is usually followed by hydrodeoxygenation (Li, Li, Yang, et al., 2013; Wang et al., 2016). The alkylation of angelica lactone and 2-MF over some acid catalysts has been reported in mild reaction conditions (Wang et al., 2016). It has become obvious that finding an excellent feedstock for the production of RD plays a key role in its sustainability and market cost, which determines its economic viability.

Numerous efforts are being made to produce RD by developing different production processes. The aim is to get RD to the market where it will be generally acceptable and used by everyone; however, on careful consideration of factors such as catalyst, feedstock, policies, technological development and a few others, this chapter focuses on understanding the challenges and opportunities in the commercialisation of RD.

4.2 Renewable diesel production route and commercialisation

Hydrogenation and decarboxylation are the two key reactions that take place in the production of RD, as described in Fig. 4.1. The process has been shown to start with removing oxygen, either as water or carbon dioxide via hydrogenation to saturate the double bonds in triglyceride. Previous authors reported the synthesis of RD using hydroxyacetone and 2-MF (Li et al., 2013).

$$C_{57}H_{104}O_6 + 6H_2 \longrightarrow 3C_{17}H_{36} + 3CO_2 + C_3H_8 \quad \text{(Decarboxylation)}$$

$$C_{57}H_{104}O_6 + 15H_2 \longrightarrow 3C_{18}H_{38} + 6H_2O + C_3H_8 \quad \text{(Hydrogenation)}$$

Figure 4.1 Decarboxylation and hydrogenation reaction in the production of renewable diesel.

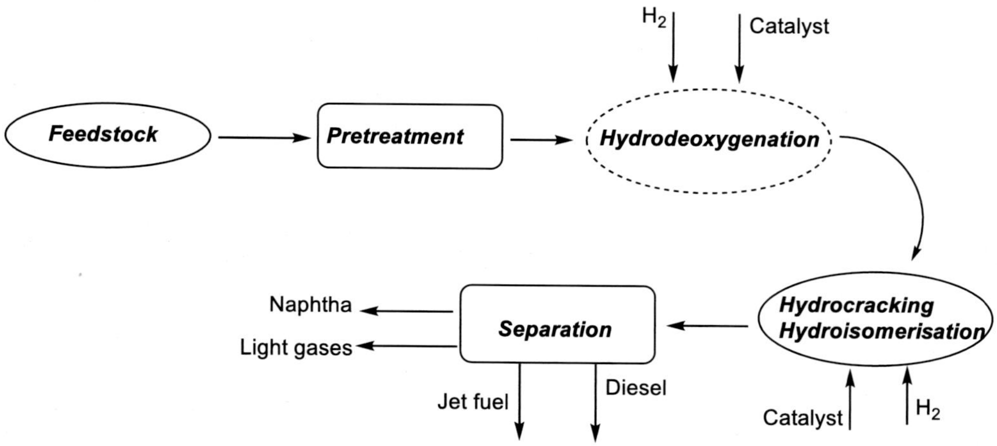

Figure 4.2 Simplified hydroprocessing pathway for renewable diesel production.

The process involved the combination of hydroxyalkylation–alkylation of 2-MF with hydroxyacetone and subsequent hydrodeoxygenation. The process revealed that the hydroxyacetone route exhibits some advantages over the conventionally known acetone route. Studies have also shown that the utilisation of biomass for the production of RD is a promising solution to the overdependence on fossil fuel energy and a means to avert the environmental challenges associated with the use of fossil fuel as this is of low toxicity, from a renewable source, with high cetane number and encouraging biodegradability (Alonso, Bond, & Dumesic, 2010; Zhou, Beltramini, Fan, & Lu, 2008). Several process pathways have been developed over time to suit the most appreciable quality of RD. However, hydroprocessing is well known as a simple process pathway. A simple description of the hydroprocessing pathway is shown in Fig. 4.2.

Commercialisation of RD has come a long way since the challenges associated with fossil fuels became a global problem. This problem is more of environmental pollution and the health hazard it brings. Commercialising RD refers to the process of transforming renewable energy technology from a laboratory or field concept into a profitable renewable energy technology that solves the current problems associated with fossil fuel energy. It is the ability of scientists, engineers and other contributors to drive their bioenergy concepts or designs into commercial success to help humanity and achieve a financial return without or with minimal risk. This path requires the scaling up laboratory or field experimental successes to industrial levels. This covers advance testing and modelling of proof of concept tested on a small scale to confirm claims or assumptions. Before scaling up, it is important to evaluate the economic viability, risk assessment and long-term impact of RD.

Interestingly, when scaleup is feasible, many intricacies or parameters come into play. This includes manufacturing complexities, technology considerations, customer expectations, production volume, cost, maintenance and sustainability. The quality of RD is very important during commercialisation. Standards have been set that the RD must meet before it can be accepted on a commercial base. In this regard, ASTM D975 and EN 590 for fuel in the United States and Europe, respectively, are well known. Different players, including accounting, engineering, law, marketing, intellectual property, etc., drive commercialisation. Experts guide the commercialisation project through the different phases before being registered for marketing.

With the growing need for RD, countries in America, Europe and Asia have gradually moved to the commercialisation stage, which still requires more effort. However, this is still very slow in Asia and African-based countries due to a lack of facilities, poverty, a low level of technological development and weak policies on green energy. Educating the government in oil-producing states in developing countries is also important to better understand the need for RD. Although several challenges, such as production cost (feedstock), process complexity (catalysis), governmental policies and technological advancement, have been identified, the need for an RD cannot be overemphasised.

The production and commercialisation of renewable fuel are growing. It is important to note that RD composes only 2.3%– 2.5% of global oil use on average (IEA, 2020). This value would have been higher if the consumption is direct, but it is mostly blended with petroleum-based fuel, which limits its use. About 71 million hectares of land are reportedly designated for biofuel production globally (Bušić et al., 2018). Regarding bioethanol, the United States, Brazil, Europe and China are the major producers. According to the Renewable Fuels Association database, the highest market for RD comes from the United

States (Mizik, 2020). The production cost is the cheapest for Brazil because of the production plant's flexibility to switch between renewable fuel production and sugar production.

On the other hand, Europe uses cereals, while the United States and China depend mostly on corn for production. The production cost for the RD comes mainly from the raw materials used (Mizik, 2020). The price of RD is expected to be cheap in countries with an abundant resource for feedstock required for its production.

4.3 Challenges and opportunities

4.3.1 Type of catalyst

Finding a cheap catalyst that is efficient and sustainable is a challenge in the production of RD. Most used catalysts produce biochar, non-condensable gases and bio-oil, which require extra processing steps for upgrade (Bharath et al., 2020). Such extra processing steps require an additional cost which increases production cost (Panwar & Paul, 2021) and, as a result, reduce the economic viability of RD. The high production cost reduces the ability of RD to compete favourably with petrol-based diesel in the market. Common catalysts used to produce RD are Fe, Ni, Ru, Pd, Rh, Pt and Co. Some of these catalysts may produce side products that complicate the process (Martinelli, Gnanamani, LeViness, Jacobs, & Shafer, 2020).

Interestingly, studies have shown that combining low-cost metal catalysts to produce bimetallic catalysts can help overcome the role of expensive individual catalysts in the production of RD. When these metal catalysts are combined, they exhibit superb properties due to their synergy, which is an improved individual metal catalyst capacity (Bhanvase & Pawade, 2018; De, Zhang, Luque, & Yan, 2016; Suryawanshi, Polagani, & Sonawane, 2018). The improved performance of the combined catalyst may be due to the change in electronic structure due to the hybridisation of the metal orbitals involved (De et al., 2016; De Coster & Poelman, 2020). The good thing about such combined metal catalysts is that they are cheap and readily available. They have high selectivity, stability and activity, yielding good-quality RD (Malins, 2021). A study reported using a sonochemically treated catalyst to perform better than conventional catalysts (Ameen et al., 2017).

Most bimetallic catalysts are produced via co-precipitation, colloidal, impregnation and solvothermal processes (Chia et al., 2021; De Coster & Poelman, 2020). Due to the ease of synthesis of these catalysts and the convenience of using them for RD synthesis new generation of combined catalysts are receiving great attention in the commercialisation of RD as these have helped reduce production cost by simplifying process mechanism (Burimsitthigul, Yoosuk, Ngamcharussrivichai, & Prasassarakich, 2021). The use of Ni–Mo and Co–Mo sulphide catalysts has been reported (Yoosuk, Sanggam, Wiengket, & Prasassarakich, 2019), which gave a conversion as high as 100%. NiMo/(Al_2O_3) and Co–Mo/Al_2O_3 have also been well-reported (Izhar, Kanesugi, Tominaga, & Nagai, 2007; Sundaramurthy, Dalai, & Adjaye, 2006). Table 4.1 shows some reported bimetallic catalysts for the production of RD. Although the use of the combined catalyst for the commercialisation of RD seems very promising, with huge opportunities to be tapped, there is a need for the exploration of this area of research, even at a large scale, to better understand the mechanism of operation as well as enhance catalytic properties and activities of the promising combined catalysts. Using biocatalysts like enzymes is subjected to several factors, making it expensive. Some of the enzymes have to be carefully stored, and when used, certain process conditions have to be maintained; if not, they become damaged. The economic cost of the enzymatic process for the production of RD is a challenge that reduces the economic feasibility. However, studies have been conducted to investigate the preparation and use of newly cultured strains of microorganisms. The process cost is still relatively high when compared to the use of inorganic catalysts.

Table 4.1 Some reported bimetallic catalyst for the synthesis of renewable diesel.

Catalyst	Synthesis method	References
Pd–Sb	Reduction–oxidation	Ye et al. (2021)
Ni–Mo	Incipient wetness	Lin, Lee, Lin, Halim, and Wang (2021)
Ir–ReOx	Sequential impregnation	Liu, Simonetti, Zheng, and Saha (2018)
Ni–Mo	Sonochemical	Ameen, Azizan, Ramli, Yusup, and Alnarabiji (2019)
Pd–Cu	Incipient wetness	Cheah et al. (2020)
CaO–La_2O_3	Impregnation	Alsultan, Asikin-Mijan, Lee, Albazzaz, and Taufiq-Yap (2017)
Pt–Re	Incipient wetness	Jin and Choi (2019)
Mo–W carbides	Wet impregnation	Tran, Akmach, and Kaliaguine (2020)

4.3.2 Type of feedstock

Among the enormous amount of waste yearly generated globally, underutilised biomass remains an exceptional source of materials used to produce RD. The most important ones are those from nonfood crops and food crops residues. Table 4.2

Table 4.2 Examples of biomass feedstock for renewable diesel production.

Class of feedstock	Feedstock
Food crops	
	Sugar beet
	Wheat
	Sugar cane
	Corn
	Sorghum
Forestry	
	Spruce wood
	Oak
	Pine
	Birch
	Bark
	Willow
Grassy crops	
	Switch grass
	Arundo donax
	Miscanthus
	Straw
Agricultural residues	
	Bagasse
	Olive tree prunings
	Grape prunings
Industrial residues	
	Hazelnut shell
	Orange peel
	Almond shell
	Olive husk
	Macroalgae
	Municipal solid waste
	Paper pulp/sludge

presents examples of biomass used as feedstock for RD production. Using both nonfood crops and food crop residues for RD production comes with many benefits in terms of energy and food security. This has found applications for agricultural wastes, including fruits and vegetables, and underutilised and lesser-known plant products. The use of agricultural waste has opened a unique window of opportunity for the energy market in developing countries where agricultural practice thrives. The ability to make use of these wastes has led to the creation of job opportunities, boosted socioeconomic growth and promoted eco-friendly products. Some wastes are by-products of primarily agricultural crops such as banana peels, cashew nutshell, coconut and rice husks, sugar cane bagasse and corn. Lignocellulosic wastes remain outstanding and can be used for RD production, as shown in Fig. 4.3. This waste, especially fruits and vegetables, which includes seed, pulp, root, bark and leaves, has helped minimise the cost of producing RD and enhance its commercialisation. A huge amount of underutilised and lesser-known crops is abandoned as waste yearly because the cost of handling and managing the waste requires an innovative solution that is not available or affordable in developing countries. This challenge has been a setback to the commercialisation of RD in this part of the world. Some countries lack the technology required to process the wastes and, as a result, discourage implementing policies to drive the commercialisation of RD.

The first-generation feedstocks are currently used in countries like the United States, Brazil and some parts of Europe (Pocha et al., 2022). The effort is presently intensified to promote the use of solid waste emanating from agro-industries. The use of palm

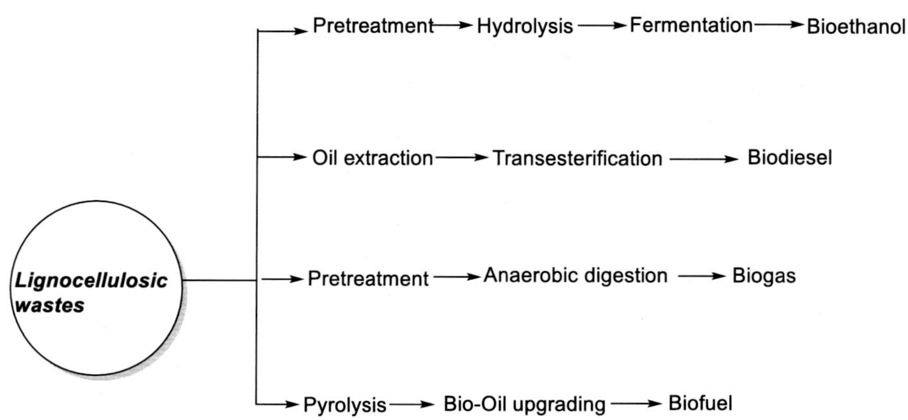

Figure 4.3 Lignocellulosic wastes as source of feedstock for renewable diesel production.

waste and the orange peel has been reported (Ahmad et al., 2021; Fazzino, Mauriello, Paone, Sidari, & Calabrò, 2021; Zhang et al., 2019). It is important to note that in most cases, the underutilised crops are processed to obtain oil or triglycerides before subjecting them to hydroprocessing to produce RD. This is a pointer that the final form of feedstock that is converted to RD is liquid oil. Therefore, as the liquid oil's content varies, the final RD's quality may also vary (Sugami, Minami, & Saka, 2016).

Furthermore, selecting a feedstock with the right properties for high-quality RD is an important step that must not be overlooked. In a recent study (Pino, Buitrago-Sierra, & López, 2020), an encouraging amount of C10 and C15, 38% and 25% were obtained from waste from oil palm industries, respectively. A study conducted on Macauba fruit was revealed to produce diesel range hydrocarbons (96.7%) and kerosene range hydrocarbons (47.3%) which is pointed to high-quality RD (Moreira, de Rezende, & Pasa, 2020). The content of the feedstock determines the type of pretreatment given and subsequent production steps that will be involved, which goes a long way in determining the cost of commercialising the RD. An example is the thermal decomposition process (such as gasification and pyrolysis) which may result in a product requiring additional upgrading steps. For the thermal decomposition process, pyrolysis, in this case, may the classified as fast, slow or flash pyrolysis depending on factors like pressure, temperature and heating rate. The final product also depends on the type of pyrolysis involved. Agricultural wastes must be pretreated before use as feedstock. The pretreatment depends on the content of the waste and, as a result, may require advanced or complex treatment, which either promotes or discourages RD commercialisation. Despite the pros and cons, the World Bioenergy Association still believes that lignocellulosic biomass is underused, and it is needed to immensely exploit its capacity for bioenergy production (Ladanai & Vinterbäck, 2009, 2010).

4.3.3 Government policy and techno-economic assessment

The commercialisation of RD still requires a detailed understanding of its economic and technical intrigues. Besides formulating policies to favour the commercialisation of RD, the sustenance and implementation of such policies are far more vital. The government does not fully implement most policies on RD commercialisation in developing countries. Key players in privately owned fossil fuel energy industries do not want the commercialisation of RD because they have invested heavily in fixed assets in the industry it will take years for them to recover such investments. This is a big challenge for local private investors in fossil fuel; the government is expected to help alleviate this challenge by providing subvention for such local investors to help them divert from non-renewable forms of energy to renewable ones (RD).

Most governments have made appreciable progress by incentivising farmers to help them grow crops for RD to avoid food scarcity. However, despite the progress made with the use of nonfood crops and food crops residue, the incentive provided by the government to encourage farmers has made some farmers concentrate on the cultivation of single-crop system specific for energy production (Jones et al., 2013; Khandaker, Qiamuddin, Majrashi, & Dalorima, 2018). This negatively impacts food security as farmers' focus is shifting away from growing food crops. The capital, operating, maintenance and feedstock costs are key factors in commercialising RD. Even when government intervention fund is available, the maintenance cost might be a challenge before the profit keeps rolling in, which may be a factor that discourages many local entrepreneurs from investing in the energy industry. In a country where there is political unrest, unstable power supply and poor access to social amenities, the operating cost might be unbearable if there is no government intervention. The economic factor is a strong key to determining the commercialisation of RD.

In the past few years, pure RD is rarely used directly; they are blended with petroleum-based diesel. There is a clear indication that policy plays a major role in the utilisation of RD, which determines the level of commercialisation. It is important to have policies that encourage higher percentage use of RD to encourage commercialisation; this may also take the form of promoting tax incentives and exemptions. Different countries practice different blending rates. E10 is commonly used in the United States, although E15 has also been approved, which increased the blending mandate to about 36 billion gallons annually (Schnepf & Yacobucci, 2010). The continuous use of RD has reduced farm prices to encourage the cultivation of important sources of feedstocks. The effort has been the sort to move to E20 and E30 (Mizik, 2020). Directive 2009/28/EC of the European Union (EU) promotes renewable energies with a 10% target on transportation (EU 2020 strategy); such an approach is good for RD, which opens windows of opportunities to be explored. The EU's regulation is based on the motive of reducing greenhouse gas emission. This action helps in the commercialiation of RD which makes it a win–win approach by providing employment as well as creating a clean environment for all.

Technological advancement is a factor. As new technology emerges, it becomes important for investors to improve and acquire them. Therefore, commercialisation depends on technology in vogue. Previous studies reported the investment cost for fast pyrolysis and hydroprocessing of biomass plant capacity of 2000 t/day to range from USD 0.6 to 1.25 L^{-1} (Anex et al., 2010; Brown, Thilakaratne, Brown, & Hu, 2013; Hu, Dang, Rover, Brown, & Wright, 2016; Shemfe, Gu, & Ranganathan, 2015; Wright, Daugaard, Satrio, & Brown, 2010), studies have also revealed that this cost range for RD is still higher than the cost of fossil diesel (Patel, Oyedun, Kumar, & Gupta, 2019). Production cost remains an undisputable factor for the commercialisation of RD. It is important to have a holistic review of the production process, which covers the development of cheap technology, finding cheap feedstock and enabling policy to help boost the production of RD in order to reduce cost, which currently hampers its commercialisation.

4.4 Future prospects

Presently, there are reports on the lab-scale hydroprocessing of mixed wood and corn stover (Bagnato, Sanna, Paone, & Catizzone, 2021; Doukeh et al., 2021; French et al., 2015; Jones et al., 2013), but there is no large-scale or commercial facility having hydroprocessing unit for bio-oil upgrading (Patel et al., 2019). This is a gap in the commercialisation of RD, which needs urgent attention. It is important to scale up the laboratory trials in order to find a more suitable approach for the industrial production of RD. The global demand for RD increased from 330×10^6 gallons in 2011 to 2.1×10^9 gallons in 2017 (Douvartzides, Charisiou, Papageridis, & Goula, 2019). On the contrary, the energy demand for 2020 was estimated to reduce by 6% globally when compared to what the demand was in 2019 (Agency, 2020; IEA, 2020; Monteiro et al., 2022), which was attributed to the effect of covid-19 (IEA, 2020; U. IEA, 2020). However, interestingly, the Global Energy Report from the IEA expected an increase in the demand for RD due to the benefits associated with its use (IEA, 2020). Affirming this estimate, a 10.3% growth in RD was reported (Ahmad & Zhang, 2020; Kemal & Genc, 2020; Newell & Raimi, 2020); this observation of an increase in demand for RD is expected to continue to rise. Therefore, the time is now for all government and private sectors to key into this opportunity to embrace RD. It is paramount that government needs to fund continuous research to support the efforts of researchers in finding the most sustainable and cheap technology for RD production while private business owners put all machinery on the ground to consolidate efforts in moving and promoting the commercialisation of RD.

The demand rate for energy in Asian countries has been shown to be high (Ahmad & Zhang, 2020), which is the case for most countries in the world. In the New Policy Scenario context, the global primary energy requirements increased by 37.0% between 2012 and 2040, which may be attributed to structural changes and energy efficiency gains in the world energy market (Ahmad & Zhang, 2020). With the rising energy changes and increase in the global economic structure rates, the average energy consumption and end-use requirements will increase. Therefore, it is expedient for countries with agricultural exports with huge waste generation to make the best use of the growing demand for energy by investing more in RD. More attention should be given to investing in large-scale production of RD and creating an enabling environment for implementing policy that drives RD commercialisation. Apart from this, there is a need to further investigate the connection between RD production and related international trade in order to understand the effect on socioeconomic development, even in developing countries with a weak economies.

4.5 Conclusion

Transitioning from fossil fuel to biofuel is vital for a sustainable clean environment. RD is an alternative energy source with a chemical composition similar to petroleum diesel. The demand for RD as an alternative form of energy for fossil fuels is increasing. Therefore, it is important to understand the challenges and opportunities for the commercialisation of RD. Process catalyst, type of feedstock, government policy and techno-economic development limit the opportunities arising from the commercialisation of RD. However, attention must be given to the upscaling of laboratory methods developed to enhance the production of RD. At the same time, government policies are enabled for adequate implementation towards the success of commercialising RD. The use of RD as a clean energy source is a green solution to the currently faced environmental problems emanating from fossil fuels.

References

Alternative Fuels Data Center. (2020). State Laws and Incentives.

Agency, I. E. (2020). *World energy outlook 2020*. OECD Publishing.

Ahmad, A., Naqvi, S. A., Jaskani, M. J., Waseem, M., Ali, E., Khan, I. A., . . . Aadil, R. M. (2021). Efficient utilization of date palm waste for the bioethanol production through Saccharomyces cerevisiae strain. *Food Science & Nutrition, 9*(4), 2066–2074.

Ahmad, T., & Zhang, D. (2020). A critical review of comparative global historical energy consumption and future demand: The story told so far. *Energy Reports, 6*, 1973–1991.

Alonso, D. M., Bond, J. Q., & Dumesic, J. A. (2010). Catalytic conversion of biomass to biofuels. *Green Chemistry, 12*(9), 1493–1513.

Alsultan, G. A., Asikin-Mijan, N., Lee, H., Albazzaz, A. S., & Taufiq-Yap, Y. (2017). Deoxygenation of waste cooking to renewable diesel over walnut shell-derived nanorode activated carbon supported CaO-La2O3 catalyst. *Energy Conversion and Management, 151*, 311–323.

Ameen, M., Azizan, M. T., Ramli, A., Yusup, S., & Alnarabiji, M. S. (2019). Catalytic hydrodeoxygenation of rubber seed oil over sonochemically synthesized Ni-Mo/γ-Al2O3 catalyst for green diesel production. *Ultrasonics Sonochemistry, 51*, 90–102.

Ameen, M., Azizan, M. T., Yusup, S., & Ramli, A. (2017). Hydroprocessing of rubber seed oil over Ni-Mo/γ-Al 2 O 3 for the green diesel production. *Chemical Engineering Transactions, 61*, 1843–1848.

Anex, R. P., Aden, A., Kazi, F. K., Fortman, J., Swanson, R. M., Wright, M. M., . . . Platon, A. (2010). Techno-economic comparison of biomass-to-transportation fuels via pyrolysis, gasification, and biochemical pathways. *Fuel, 89*, S29–S35.

Bagnato, G., Sanna, A., Paone, E., & Catizzone, E. (2021). Recent catalytic advances in hydrotreatment processes of pyrolysis bio-oil. *Catalysts, 11*(2), 157.

Bhanvase, B. A., & Pawade, V. B. (2018). Advanced nanomaterials for green energy: Current status and future perspectives. *Nanomaterials for Green Energy* (pp. 457–472). Elsevier.

Bharath, G., Hai, A., Rambabu, K., Banat, F., Jayaraman, R., Taher, H., . . . Schmidt, J. E. (2020). Systematic production and characterization of pyrolysis-oil from date tree wastes for bio-fuel applications. *Biomass and Bioenergy, 135*, 105523.

Bharath, G., Rambabu, K., Hai, A., Morajkar, P. P., Salkar, A. V., Hasan, S. W., . . . Banat, F. (2021). Highly selective etherification of fructose and 5-hydroxymethylfurfural over a novel Pd-Ru/MXene catalyst for sustainable liquid fuel production. *International Journal of Energy Research, 45*(10), 14680–14691.

Brown, T. R., Thilakaratne, R., Brown, R. C., & Hu, G. (2013). Techno-economic analysis of biomass to transportation fuels and electricity via fast pyrolysis and hydroprocessing. *Fuel, 106*, 463–469.

Burimsitthigul, T., Yoosuk, B., Ngamcharussrivichai, C., & Prasassarakich, P. (2021). Hydrocarbon biofuel from hydrotreating of palm oil over unsupported Ni–Mo sulfide catalysts. *Renewable Energy, 163*, 1648–1659.

Bušić, A., Kundas, S., Morzak, G., Belskaya, H., Maretko, N., Šantek, M. I., . . . Šantek, B. (2018). Recent trends in biodiesel and biogas production. *Food technology and biotechnology, 56*(2), 152.

Cheah, K. W., Taylor, M. J., Osatiashtiani, A., Beaumont, S. K., Nowakowski, D. J., Yusup, S., . . . Kyriakou, G. (2020). Monometallic and bimetallic catalysts based on Pd, Cu and Ni for hydrogen transfer deoxygenation of a prototypical fatty acid to diesel range hydrocarbons. *Catalysis Today, 355*, 882–892.

Chia, S. R., Nomanbhay, S., Ong, M. Y., Chew, K. W., Khoo, K. S., Karimi-Maleh, H., & Show, P. L. (2021). Recent Development of Renewable Diesel Production Using Bimetallic Catalysts. *Frontiers in Energy Research*, 621.

De, S., Zhang, J., Luque, R., & Yan, N. (2016). Ni-based bimetallic heterogeneous catalysts for energy and environmental applications. *Energy & Environmental Science*, *9*(11), 3314–3347.

De Coster, V., & Poelman, H. (2020). Catalyst design via colloidal synthesis. In *Encyclopedia*.

Doukeh, R., Bombos, D., Bombos, M., Oprescu, E.-E., Dumitrascu, G., Vasilievici, G., & Calin, C. (2021). Catalytic hydrotreating of bio-oil and evaluation of main noxious emissions of gaseous phase. *Scientific Reports*, *11*(1), 1–13.

Douvartzides, S. L., Charisiou, N. D., Papageridis, K. N., & Goula, M. A. (2019). Green diesel: Biomass feedstocks, production technologies, catalytic research, fuel properties and performance in compression ignition internal combustion engines. *Energies*, *12*(5), 809.

Fazzino, F., Mauriello, F., Paone, E., Sidari, R., & Calabrò, P. S. (2021). Integral valorization of orange peel waste through optimized ensiling: Lactic acid and bioethanol production. *Chemosphere*, *271*, 129602.

French, R. J., Black, S. K., Myers, M., Stunkel, J., Gjersing, E., & Iisa, K. (2015). Hydrotreating the organic fraction of biomass pyrolysis oil to a refinery intermediate. *Energy & Fuels*, *29*(12), 7985–7992.

Han, J., Sen, S. M., Alonso, D. M., Dumesic, J. A., & Maravelias, C. T. (2014). A strategy for the simultaneous catalytic conversion of hemicellulose and cellulose from lignocellulosic biomass to liquid transportation fuels. *Green Chemistry*, *16*(2), 653–661.

Hu, W., Dang, Q., Rover, M., Brown, R. C., & Wright, M. M. (2016). Comparative techno-economic analysis of advanced biofuels, biochemicals, and hydrocarbon chemicals via the fast pyrolysis platform. *Biofuels*, *7*(1), 57–67.

IEA. (2020). Oil Market Report—April 2020. In: International Energy Agency Paris, France.

IEA, U. (2020). Global energy review 2020. *Ukraine. [Online]* https://www.iea.org/countries/ukraine [Accessed: 2020-09-10].

i Nogué, V. S., Black, B. A., Kruger, J. S., Singer, C. A., Ramirez, K. J., Reed, M. L., ... Yeap, R. Y. (2018). Integrated diesel production from lignocellulosic sugars via oleaginous yeast. *Green Chemistry*, *20*(18), 4349–4365.

Izhar, S., Kanesugi, H., Tominaga, H., & Nagai, M. (2007). Cobalt molybdenum carbides: Surface properties and reactivity for methane decomposition. *Applied Catalysis A: General*, *317*(1), 82–90.

Jin, M., & Choi, M. (2019). Hydrothermal deoxygenation of triglycerides over carbon-supported bimetallic PtRe catalysts without an external hydrogen source. *Molecular Catalysis*, *474*, 110419.

Jones, S., Meyer, P., Snowden-Swan, L., Padmaperuma, A., Tan, E., Dutta, A., ... Cafferty, K. (2013). *Process design and economics for the conversion of lignocellulosic biomass to hydrocarbon fuels: fast pyrolysis and hydrotreating bio-oil pathway.*

Kemal, K., & Genc, M. S. (2020). Effects of 2019 Novel Coronavirus (COVID-19) Outbreak on global energy demand and the electricity production with renewables: a comprehensive survey. *Sigma Journal of Engineering and Natural Sciences*, *38*(3), 1369–1380.

Khandaker, M. M., Qiamuddin, K., Majrashi, A., & Dalorima, T. (2018). Bio-ethanol production from fruit and vegetable waste by using saccharomyces cerevisiae. *Bioethanol Technologies*.

Ladanai, S., & Vinterbäck, J. (2009). Global potential of sustainable biomass for energy.

Ladanai, S., & Vinterbäck, J. (2010). Certification criteria for sustainable biomass for energy.

Li, G., Li, N., Li, S., Wang, A., Cong, Y., Wang, X., & Zhang, T. (2013). Synthesis of renewable diesel with hydroxyacetone and 2-methyl-furan. *Chemical Communications, 49*(51), 5727–5729.

Li, G., Li, N., Yang, J., Wang, A., Wang, X., Cong, Y., & Zhang, T. (2013). Synthesis of renewable diesel with the 2-methylfuran, butanal and acetone derived from lignocellulose. *Bioresource Technology, 134*, 66–72.

Lin, J. K., Lee, C. P., Lin, H. P., Halim, E., & Wang, W. C. (2021). The production of hydro-processed renewable diesel over the nonsulfide catalyst. *International Journal of Energy Research, 45*(13), 19043–19061.

Liu, S., Simonetti, T., Zheng, W., & Saha, B. (2018). Selective Hydrodeoxygenation of Vegetable Oils and Waste Cooking Oils to Green Diesel Using a Silica-Supported Ir–ReOx Bimetallic Catalyst. *ChemSusChem, 11*(9), 1446–1454.

Low, S. S., Bong, K. X., Mubashir, M., Cheng, C. K., Lam, M. K., Lim, J. W., ... Show, P. L. (2021). Microalgae cultivation in palm oil mill effluent (POME) treatment and biofuel production. *Sustainability, 13*(6), 3247.

Malins, K. (2021). Synthesis of renewable hydrocarbons from vegetable oil feedstock by hydrotreatment over selective sulfur-free SiO2-Al2O3 supported monometallic Pd, Pt, Ru, Ni, Mo and bimetallic NiMo catalysts. *Fuel, 285*, 119129.

Martinelli, M., Gnanamani, M. K., LeViness, S., Jacobs, G., & Shafer, W. D. (2020). An overview of Fischer-Tropsch Synthesis: XtL processes, catalysts and reactors. *Applied Catalysis A: General, 608*, 117740.

Mizik, T. (2020). Impacts of international commodity trade on conventional biofuels production. *Sustainability, 12*(7), 2626.

Monteiro, R. R., dos Santos, I. A., Arcanjo, M. R., Cavalcante, C. L., Jr, de Luna, F. M., Fernandez-Lafuente, R., & Vieira, R. S. (2022). Production of Jet Biofuels by Catalytic Hydroprocessing of Esters and Fatty Acids: A Review. *Catalysts, 12*(2), 237.

Moreira, J. d. B. D., de Rezende, D. B., & Pasa, V. M. D. (2020). Deoxygenation of Macauba acid oil over Co-based catalyst supported on activated biochar from Macauba endocarp: A potential and sustainable route for green diesel and biokerosene production. *Fuel, 269*, 117253.

Newell, R. G., & Raimi, D. (2020). Global energy outlook comparison methods: 2020 update. *Resources for the Future: Washington, DC, USA.*

Nakagawa, Y., Tamura, M., & Tomishige, K. (2013). Catalytic reduction of biomass-derived furanic compounds with hydrogen. *ACS Catalysis, 3*(12), 2655–2668.

Okeke, I. J., Sahoo, K., Kaliyan, N., & Mani, S. (2020). Life cycle assessment of renewable diesel production via anaerobic digestion and Fischer-Tropsch synthesis from miscanthus grown in strip-mined soils. *Journal of Cleaner Production, 249*, 119358.

Oladipupo Kareem, M., Edathil, A. A., Rambabu, K., Bharath, G., Banat, F., Nirmala, G., & Sathiyanarayanan, K. (2021). Extraction, characterization and optimization of high quality bio-oil derived from waste date seeds. *Chemical Engineering Communications, 208*(6), 801–811.

Panwar, N. L., & Paul, A. S. (2021). An overview of recent development in bio-oil upgrading and separation techniques. *Environ. Eng. Res, 26*(5), 200382.

Patel, M., Oyedun, A. O., Kumar, A., & Gupta, R. (2019). A techno-economic assessment of renewable diesel and gasoline production from aspen hardwood. *Waste and Biomass Valorization, 10*(10), 2745–2760.

Pino, N., Buitrago-Sierra, R., & López, D. (2020). Conversion of Biomass-Derived Furanics to Fuel-Range Hydrocarbons: Use of Palm Oil Empty Fruit Bunches. *Waste and Biomass Valorization, 11*(2), 565–577.

Pocha, C. K. R., Chia, S. R., Chia, W. Y., Koyande, A. K., Nomanbhay, S., & Chew, K. W. (2022). Utilization of agricultural lignocellulosic wastes for biofuels and green diesel production. *Chemosphere, 290*, 133246.

Schnepf, R. D., & Yacobucci, B. D. (2010). *Renewable fuel standard (RFS): overview and issues* (Vol. 40155). Congressional Research Service Washington, DC.

Shemfe, M. B., Gu, S., & Ranganathan, P. (2015). Techno-economic performance analysis of biofuel production and miniature electric power generation from biomass fast pyrolysis and bio-oil upgrading. *Fuel, 143*, 361–372.

Sugami, Y., Minami, E., & Saka, S. (2016). Renewable diesel production from rapeseed oil with hydrothermal hydrogenation and subsequent decarboxylation. *Fuel, 166*, 376–381.

Sundaramurthy, V., Dalai, A., & Adjaye, J. (2006). Comparison of P-containing γ-Al2O3 supported Ni-Mo bimetallic carbide, nitride and sulfide catalysts for HDN and HDS of gas oils derived from Athabasca bitumen. *Applied Catalysis A: General, 311*, 155–163.

Suryawanshi, P. L., Polagani, R. K., & Sonawane, S. H. (2018). *Synthesis of Nanomaterials for Fuel Cell Applications. Nanomaterials for Green Energy* (pp. 205–226). Elsevier.

Tran, C.-C., Akmach, D., & Kaliaguine, S. (2020). Hydrodeoxygenation of vegetable oils over biochar supported bimetallic carbides for producing renewable diesel under mild conditions. *Green Chemistry, 22*(19), 6424–6436.

Veriansyah, B., Han, J. Y., Kim, S. K., Hong, S.-A., Kim, Y. J., Lim, J. S., ... Kim, J. (2012). Production of renewable diesel by hydroprocessing of soybean oil: Effect of catalysts. *Fuel, 94*, 578–585.

Wang, W., Li, N., Li, S., Li, G., Chen, F., Sheng, X., ... Zhang, T. (2016). Synthesis of renewable diesel with 2-methylfuran and angelica lactone derived from carbohydrates. *Green Chemistry, 18*(5), 1218–1223.

Wright, M. M., Daugaard, D. E., Satrio, J. A., & Brown, R. C. (2010). Techno-economic analysis of biomass fast pyrolysis to transportation fuels. *Fuel, 89*, S2–S10.

Ye, C., Peng, M., Cui, T., Tang, X., Wang, D., Jiao, M., ... Li, Y. (2021). Revealing the surface atomic arrangement of noble metal alkane dehydrogenation catalysts by a stepwise reduction-oxidation approach. *Nano Research*, 1–7.

Yoosuk, B., Sanggam, P., Wiengket, S., & Prasassarakich, P. (2019). Hydrodeoxygenation of oleic acid and palmitic acid to hydrocarbon-like biofuel over unsupported Ni-Mo and Co-Mo sulfide catalysts. *Renewable Energy, 139*, 1391–1399.

Zhang, Y., Cui, Y., Chen, P., Liu, S., Zhou, N., Ding, K., ... Cheng, Y. (2019). Gasification technologies and their energy potentials. *Sustainable resource recovery and zero waste approaches* (pp. 193–206). Elsevier.

Zhou, C.-H. C., Beltramini, J. N., Fan, Y.-X., & Lu, G. M. (2008). Chemoselective catalytic conversion of glycerol as a biorenewable source to valuable commodity chemicals. *Chemical Society Reviews, 37*(3), 527–549.

5

Current status and future outlook for renewable diesel industry

Veena Singh
Department of Material Science and Engineering, KTH-Royal Institute of Technology, Stockholm, Sweden

5.1 Introduction

Renewable diesel is among the potential contributors towards expanding renewable energy in the transport sector. Over the past 10 years, the renewable industry and the world have changed more than we could have ever anticipated. There are several ongoing investigations for production and process improvements for better performance. These can contribute towards a sustainable society if applied in daily life. Perhaps, this alternative renewable diesel is capable and a potential candidate to replace conventional fossil fuel better and more sustainably (Aatola, Larmi, Sarjovaara, & Mikkonen, 2008; Rapier, 2008). This chapter gives an overview of global energy crises due to increasing GHG emissions and global warming with respect to different energy-consuming sectors. The current biofuel scenario in terms of ethanol, biodiesel and renewable diesel has been explained in addition to market trends for biofuel production and consumption.

Furthermore, the difference between renewable diesel and biodiesel has been summarised. Fuel standards and regulations applicable to different renewable fuels have been identified and discussed. This chapter reports commercial renewable diesel producers and overall installed and planned renewable diesel plants. In this chapter, the worldwide potential production capacity of different countries for renewable diesel has been reviewed. This chapter provides perceptions for the future growth area in the renewable diesel sector.

Renewable Diesel. DOI: https://doi.org/10.1016/B978-0-323-91153-5.00005-4

5.2 Global energy crises

If we could define our restrictions for fossil fuel usage and a roadmap to deal with the current crisis, it would be crucial to avoid downfall and secure our thriving future. One of the most recognisable and well-known boundaries is the rising global temperature, which has increased yearly since growing urbanisation. This is one of the critical concerns, and perhaps we have already crossed the boundary, which can be observed in today's climate change, including the inevitable melting of glaciers. With such a huge global warming increase, the risk of crossing the tipping point also increases. The tipping point is the point from which there is no option of a possible return to normalcy. The Earth's average temperature and CO_2 concentration in the atmosphere were traceable thousands of years before. However, with the industrial revolution, this concentration changed drastically, and during 1988, we crossed 350 ppm concentration for CO_2 emission. This was the first moment we crossed that boundary, and ever since, we have been at threat of provoking the fluctuations that lead to runway warnings. We have reached approximately 415 ppm concentration of CO_2 in the air, which is towards the danger zone in the climate boundary. We are approaching 450 ppm CO_2 emission concentration speedily, a predefined global borderline danger zone through the uncertainty range. Suppose we exceed this limit and enter into the high-risk zone. In that case, that will be irreversible tipping points and become inevitable, and this is a conventional estimate, given that signs of tipping points are all around us now. Climate planetary boundary equals 1.5°C warming; if we allow ourselves to go beyond 1.5°C, that will be a huge risk. Currently, we are at 1.1°C and rapidly moving towards 1.5°C, and if we want to stay within the planetary boundary of climate, then we need to reach a fossil fuel–free world economy within the next 30 years (The Intergovernmental Panel on Climate Change IPCC, 2022).

The planetary boundaries have given us a clear path ahead. Simply choosing renewable energy could transform our future on Earth and improve our lives. That ultimately would reduce the risks of conflict and instability. From the data obtained, the coming time will be the decisive decade for humanity's future. It is a remarkable time to be alive, but it also carries great responsibility to act decisively. Focus on problem-solving will be crucial to overcoming the conflicts raised due to the excessive use of fossil fuels and their negative impact on the atmosphere (Nomanbhay, Hussein, & Ong, 2018). The Intergovernmental Panel on Climate

Change has reported that the most significant factor for global warming is emission from fossil fuels (Fossil fuels and climate change: The facts, 2022). Fossil fuel burning releases a huge amount of GHG gas into the atmosphere. These GHGs trap heat in the atmosphere and contribute to global warming.

Globally, GHG emissions have grown drastically, and the period between 1990 and 2018 it was almost 50% of increment (IEA, 2020). Fig. 5.1 illustrates the main sectors responsible for GHG emissions, including the energy, industry, waste and agriculture sector.

Energy consumption is one of the biggest contributors to GHG emissions, and more than 70% contribution is from the energy sector. Other sectors are agriculture which contributes around 18%, and the waste sector contributes approximately 3%, whereas the industrial sector contributes more than 5% of GHG emissions (IEA, 2020).

More detailed contributions and share of GHG emission through sub-sectors is shown in Fig. 5.2. The energy sector mostly includes transport, energy in buildings for electricity and heat, energy in industry, energy in agriculture and fishing, unallocated fuel combustion and fugitive emission through the energy sector's contribution to GHG emission. Within the energy sector, energy in industry is responsible for the highest emission of ~24%, followed by electricity and heat generation,

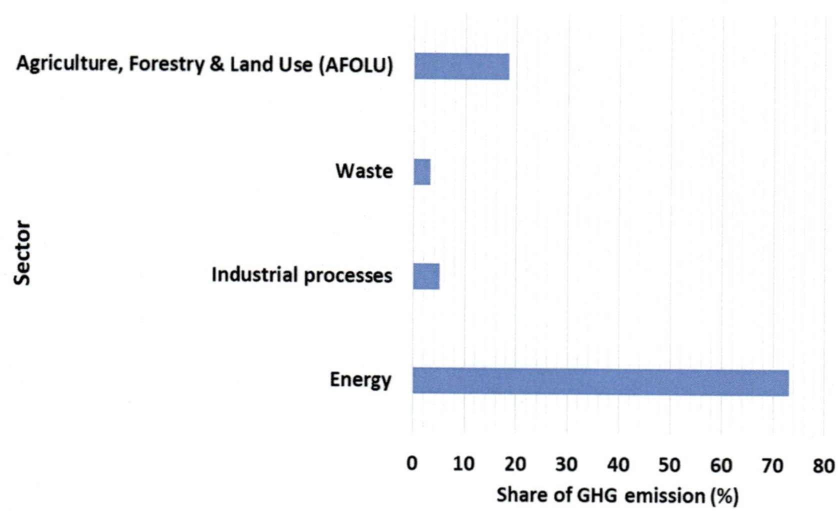

Figure 5.1 Share of GHG emission from different sectors. Data from International Energy Agency. (2020). Modified by WRI. Retrieved April 3, 2022, from http://www.iea.org/statistics.

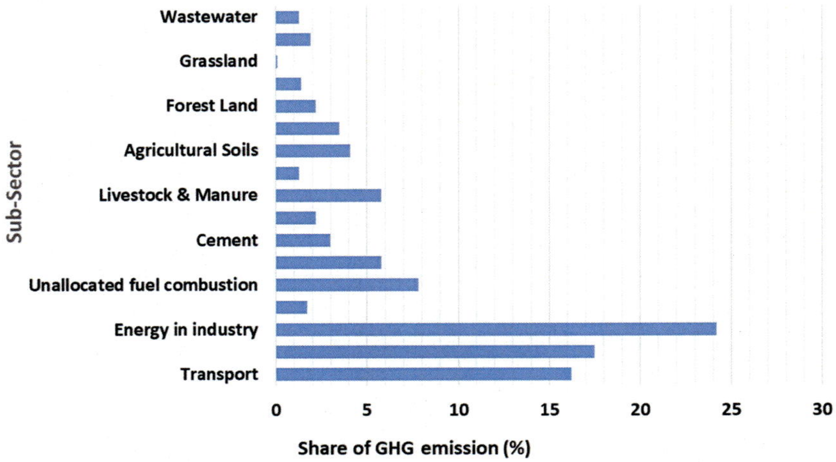

Figure 5.2 Share of GHG emission from sub-sectors. Data from International Energy Agency. (2020). Modified by WRI. Retrieved April 3, 2022, from http://www.iea.org/statistics.

which is ~17% of emissions, and transport leads to ~16% of GHG emissions (IEA, 2020).

Cement, chemical and petrochemical are the sub-sectors of an industrial process for GHG emission. Agriculture, forestry and land use have been subdivided into livestock manure, cultivation, agriculture soil, crop burning, forestland, cropland and grassland. At the same time, waste emission includes wastewater and landfill sources for GHG emission.

The fastest growing sectors for GHG emission are the industrial process, which grew by 87%; transportation grew by 79% and the manufacturing industry by 56% since 1990. This GHG emission contribution is not only from CO_2 emission but also from excessive air conditioning and refrigeration usage. That activity ultimately produces hydrofluorocarbon, one of the potent GHG gases (Climate Watch, 2022).

Data for worldwide fossil CO_2 emissions from 1994 to 2020 is shown in Fig. 5.3 (Friedlingstein et al., 2021).

All values are in a million tons of carbon per year, and for million tons of CO_2 per year, values are multiplied by 3.664 (1 MtC = 1 million tons of carbon = 3.664 million tonnes of CO_2). CO_2 emissions have increased rapidly in the past few years. We can observe from the graph that, during 2020, CO_2 emissions declined by 5.8% (9500 million tonnes of carbon), the largest decline due to the COVID-19 pandemic that had a huge impact on the industry and transport sector (IEA, 2021). Considering fossil CO_2 emissions and GHG emissions in the

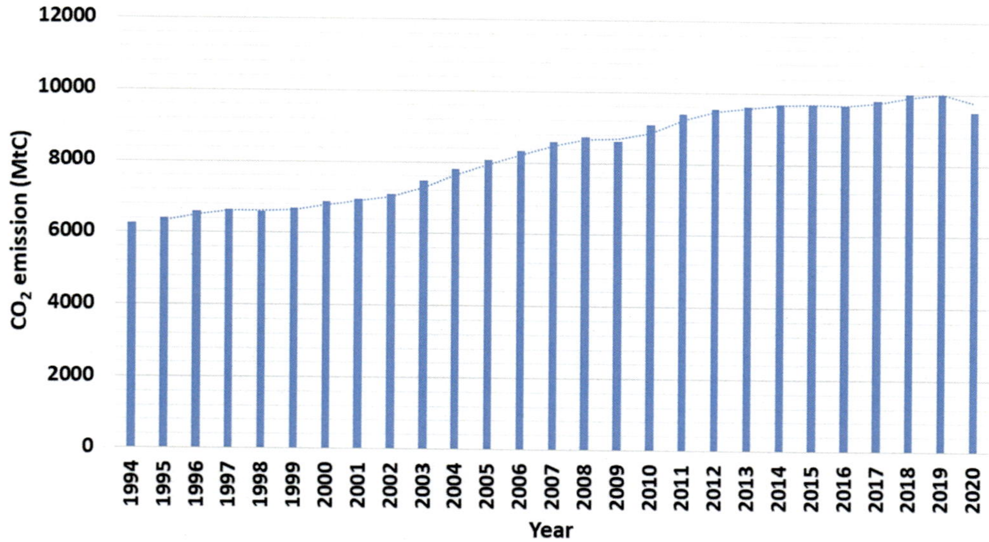

Figure 5.3 Worldwide fossil CO_2 emissions (Friedlingstein et al., 2021).

transport sector, finding a better alternative to fossil fuel is crucial.

5.3 Current scenario for biofuel

Sustainable and green approaches are appealing reflections to preserve the atmosphere. Since major contributor to GHG emissions is the transport sector, which mainly uses petroleum fuel, natural gas or biofuels (U.S. Energy Information Administration, 2021a). Petroleum products derived from petroleum or treatment of natural gas are also called secondary fuel, for example, gasoline, diesel, jet fuel, residual fuel, etc. In contrast, biofuels are further categorised into ethanol, biodiesel and renewable diesel obtained from the processing of the renewable feedstock. Each fuel has a different application based on the engine type, way of transport and application for other sectors. Due to major climate disasters and the limited availability of conventional petroleum fuels, biofuels are the key player and can replace fossil fuels. The application of biofuels, including ethanol, biodiesel and renewable diesel, for different sectors, is shown in Fig. 5.4 (Integrated Flow Solutions, 2020; U.S. Department of Energy, 2021; U.S. Energy Information Administration, 2021b).

The worldwide annual biofuel production report is shown in Fig. 5.5. The renewable energy market has reported annual

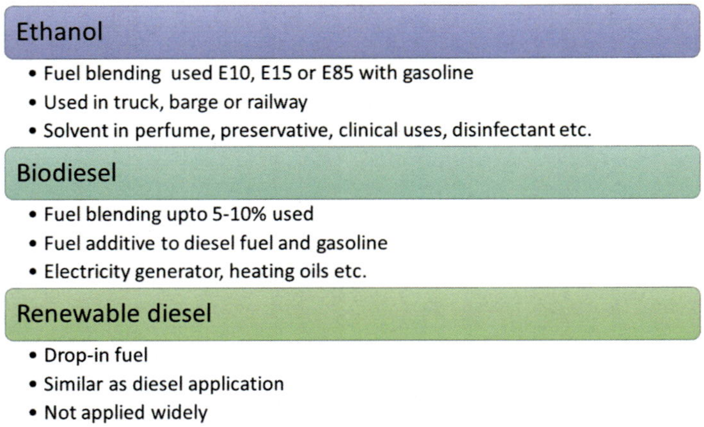

Figure 5.4 Biofuel application in different sectors. From Integrated Flow Solutions (2020). Biodiesel guide—Sources, production, uses, & regulations. Retrieved April 3, 2022, from https://ifsolutions.com/how-is-biodieselmade-sources-uses/; U.S. Department of Energy. (2021). Ethanol fuel basics. Retrieved April 5, 2022, from https://afdc.energy.gov/fuels/ethanol_fuel_basics.html; and U.S. Energy Information Administration. (2021b). Use of energy, transportation explained. Retrieved April 5, 2022, from https://www.eia.gov/energyexplained/use-of-energy/transportation.php.

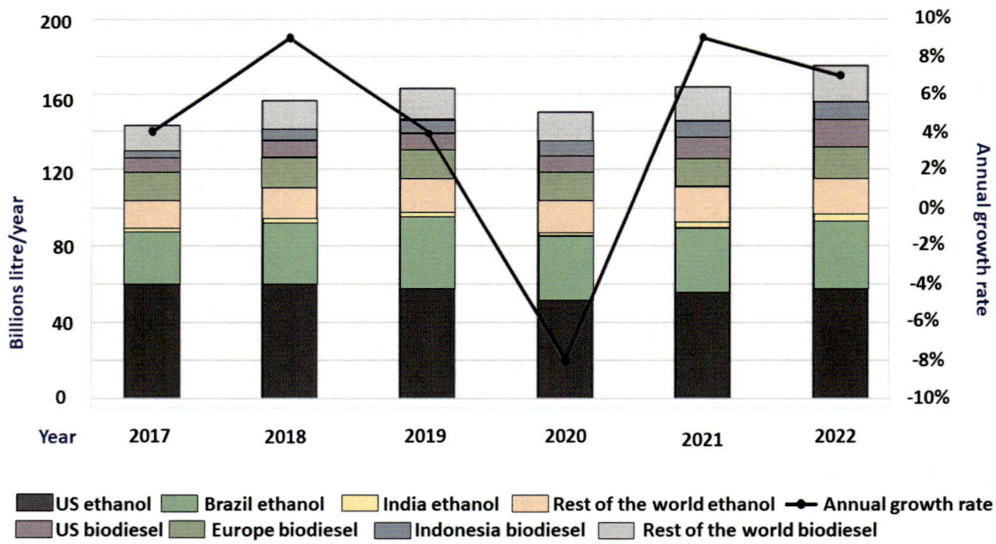

Figure 5.5 Annual biofuel production by country from 2017 to 2022. From International Energy Agency. (2021). *Renewable energy market update: Outlook for 2021 and 2022.* Retrieved April 5, 2022, from https://iea.blob.core.windows.net/assets/18a6041d-bf13-4667-a4c2-8fc008974008/RenewableEnergyMarketUpdate-Outlookfor2021and2022.pdf.

biofuel production from 2017 to 2022 for the transport sector (International Energy Agency, 2021). The annual growth rate shows an increase from 2017 to 2019 and a decrease in 2020. This unexpected decrease was reported due to the ongoing

COVID-19 crisis. On the other hand, production capacity is anticipated to witness approximately a 9% growth rate during 2021 because of the substantial capacity extension through new projects. Renewable energy sources have been intensively studied and consumed to accomplish the energy supply and enormous demand for biofuels, despite COVID-19.

Worldwide biofuel production decreased ~6% during 2020 and was 1677 thousand barrels equivalent per day (tbe/day). However, during 2019 production was 1790 tbe/day. Biofuel production data worldwide is shown in Table 5.1. The table represents the production growth rate for 2020 and 2009–19 and biofuel production in tbe/day for 2020, 2019 and 2018 (Statistical Review of World Energy, 2021).

The global biofuel consumption decreased ~5% during 2020 and was 1679 tbe/day. On the other hand, during 2019, consumption was 1772 tbe/day. Biofuel consumption data is shown in Table 5.2. The table represents the annual biofuel consumption growth rate for 2020 and 2009–19 and biofuel consumption in tbe/day for 2020, 2019 and 2018 (Statistical Review of World Energy, 2021).

If fuel demand for the transport sector and policy support in the global market continues to increase, biofuel production could reach 162 billion litres during 2021 and 2022, which will be close to 2019 production. Due to the pandemic in 2020 this biofuel production was affected and lower than in 2019. Global biofuel production in renewable diesel, ethanol and biodiesel is

Table 5.1 Worldwide biofuel production.

Country	Growth rate per annum		Production (tbe/day)		
	2020	2009–19	2020	2019	2018
North America	–9.6%	4.7%	623	689	707
South and Central America	–7.5%	5.1%	446	482	461
Europe	–4.6%	5.6%	295	309	304
Middle East	–	2.7%	<0.5	<0.5	<0.5
Africa	–	–6.0%	3	3	3
CIS	–	–3.7%	<0.5	<0.5	<0.5
Asia Pacific	1.1%	17.1%	310	306	246
Total	–6.3%	6.2%	1677	1790	1721

Source: *From Statistical Review of World Energy. (2021).* BP energy outlook *(70th ed.). Retrieved April 5, 2022, from https://www.bp.com/en/global/corporate/energy-economics/statistical-review-of-world-energy.htm.*

Table 5.2 Worldwide biofuel consumption.

Country	Growth rate per annum		Consumption (tbe/day)		
	2020	2009–19	2020	2019	2018
North America	−11.0%	4.5%	598	671	669
South and Central America	−3.8%	6.5%	464	482	439
Europe	1.0%	5.2%	339	336	323
Middle East	−26.4%	−4.4%	<0.5	<0.5	<0.5
Africa	−0.9%	8.9%	<0.5	<0.5	<0.5
CIS	12.6%	21.9%	3	3	2
Asia Pacific	−1.6%	14.0%	274	279	223
Total	−5.2%	6.2%	1679	1772	1658

Source: *From Statistical Review of World Energy. (2021).* BP energy outlook *(70th ed.). Retrieved April 5, 2022, from https://www.bp.com/en/global/corporate/energy-economics/statistical-review-of-world-energy.htm.*

Table 5.3 Global biofuel production for 2019 and forecast up to 2025.

Year	Renewable diesel (billion litres)	Biodiesel (billion litres)	Ethanol (billion litres)
2019	7	41	115
2020	9	37	98
2021	10	43	109
2022	12	45	114
2023–25	17	46	119

Source: *From International Energy Agency. (2020b).* IEA renewables. *Retrieved April 5, 2022, from https://www.iea.org/reports/renewables-2020/transport-biofuels.*

displayed in Table 5.3 (International Energy Agency IEA Renewables, 2020). Output for 2022 is anticipated to increase to 171 billion litres. Increasing fuel demand and biofuel policies support in American and Asian countries allow markets for additional growth. During 2023–25, the anticipated global biofuel output is 182 billion litres, with the highest increase in ethanol production in China and Brazil. Biodiesel and renewable diesel production are expected to be highest in the United States and Asian regions, respectively.

Biofuels are at the centre of attraction as an alternative fuel for the transport sector due to lower carbon emissions and to attain energy security through local supplies. Biofuel demand for the transport sector declined by 8% during 2019–20 and was 150 billion litres. Renewable diesel production was 9 billion litres in 2020; on the other hand, it was 7 billion litres in 2019. The largest decline in biofuel production output was observed in Brazil, the United States and Europe, and it accounted for 90% of the market during 2019–20. This decline was due to the Covid-19 pandemic, affecting biofuel production industries immensely. Whereas global biofuel production is expected to recover during 2021, according to the forecast, this recovery was not even. Biofuels are expected to meet around 5.4% of road transport energy demand by 2025, increasing from 4.8% in 2019 (International Energy Agency IEA Renewables, 2020). Biofuel production in terms of transport increases globally, as shown in Fig. 5.6. Ethanol production has had a huge impact on the transportation sector. However, renewable diesel and biodiesel are also increasing rapidly.

5.4 Renewable diesel versus biodiesel

Renewable diesel and biodiesel have witnessed growing interest due to low carbon emission and sustainability perspectives. Even though both have similar properties in feedstock, less carbon emission and low impact on climate, they have some significant

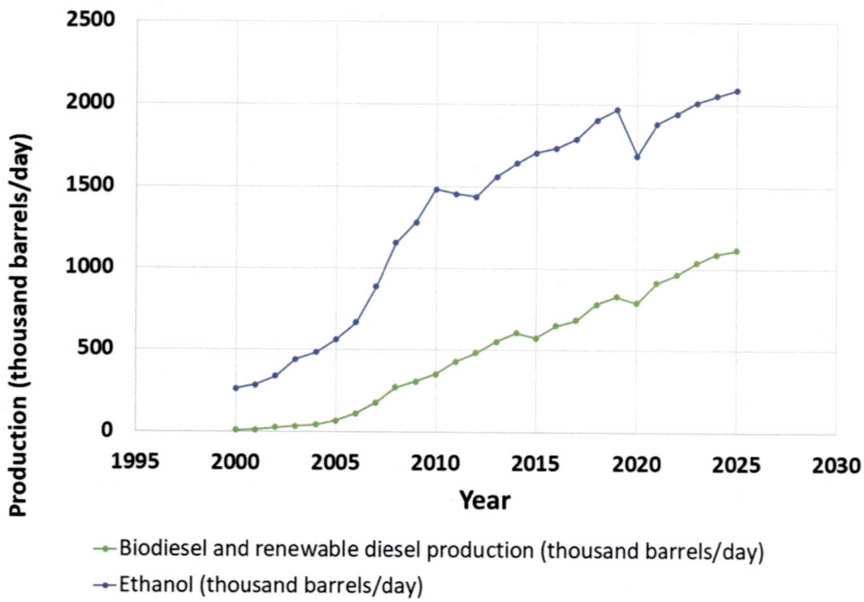

Figure 5.6 Worldwide biofuel production forecast for transport from 2000 to 2025. From International Energy Agency. (2020). *IEA renewables*. Retrieved April 5, 2022, from https://www.iea.org/reports/renewables-2020/transport-biofuels.

Table 5.4 Comparison between biodiesel and renewable diesel (Agwu, Valera-Medina, Katrašnik, & Seljak, 2021).

Biodiesel	Renewable diesel
• Biodiesel can be domestically produced, which is renewable and biodegradable. It has a higher cloud point compared to conventional diesel, so blending with petroleum diesel is a must for application in a diesel engine. • Transesterification reaction is required for biodiesel production. • In the transesterification process, glycerol is the byproduct that has multiple applications in various sectors.	• Chemical properties of renewable diesel are similar to petroleum diesel, and hence no need to blend before application. It can be used in existing petroleum pipelines, storage tanks and diesel engines. • Thermochemical process is required for renewable diesel production. • Renewable naphtha and renewable propane are the byproducts generated from renewable diesel production and can be used as it is or to reduce carbon footprint in the production process.

differences. Renewable diesel differs from biodiesel in production, blending % and fuel properties (Gebremariam and Marchetti, 2018; Michigan farm news report, 2022). Table 5.4 presents a comparison between biodiesel and renewable diesel.

5.5 Why renewable diesel?

Technical advancement, increase in transportation, industrialisation, agriculture, infrastructure and population are the major factors responsible for the terrific rise in global energy demand in recent years.

For substituting conventional petroleum diesel, diesel fuels such as biodiesel and renewable diesel are the potential candidates. Biodiesel is also represented as FAME, that is fatty acid methyl ester. It has advantages such as less smoke, reduction of GHG gases, no sulphur, less particulate matter emission and complete fuel combustion due to excess oxygen content compared to fossil fuel (Gebremariam & Marchetti, 2018). However, the major drawbacks of FAME are autoxidation and instability towards long-term storage at room temperature are the foremost concerns since they are prone to oxidative degradation because of unsaturated fatty acids (Karavalakis, Hilari, Givalou, Karonis, & Stournas, 2011). This ultimately causes filter clogging in diesel engines, fuel thickening and sediment formation.

On the other hand, renewable diesel is a better substitute to overcome biodiesel's drawbacks and provide cleaner combustion than diesel fuel. The feedstock for renewable diesel production is similar to biodiesel, such as vegetable oils, cooking oil, animal fats and lignocellulose biomass (Chia et al., 2022). Renewable diesel can be stored longer due to its properties similar to petroleum diesel. Renewable diesel is a mixture of straight and branched saturated hydrocarbons (C15–C18) without oxygen (Douvartzides, Charisiou, Papageridis, & Goula, 2019).

Renewable or green diesel, derived through biomass hydro treatment, is entirely suitable for diesel engine use. In the presence of a catalyst, the feedstock is treated by different routes such as gasification, pyrolysis, biochemical/thermochemical or hydro treatment. Schematic representation of renewable diesel production process is shown in Fig. 5.7. Moreover, renewable diesel has several advantages, including easy blending, high performance, excellent storage stability, high heating value, sulphur-free, non-corrosive, easy to use and 100% renewable (Douvartzides et al., 2019). These properties make renewable diesel fuel more attractive in the present time.

Aromatic and impure compounds are absent in renewable diesel compared to fossil fuels. That eventually affects fuel combustion properties, which improves fuel efficiency. A high cetane number contributes to clean and efficient combustion, which keeps the engine cleaner than traditional biodiesel. Due to the high cetane number, renewable diesel also improves combustion in cold weather (up to −34°C). Renewable diesel can be stored without affecting or deteriorating fuel quality for extended periods due to its stability. Renewable diesel perhaps has the potential to compete with conventional diesel and increase profit margins through blending. As fuel properties are considerably similar to diesel and considered a "Drop-In" fuel, it is compatible with current engine systems. Improvement in gaseous emission from combustion and

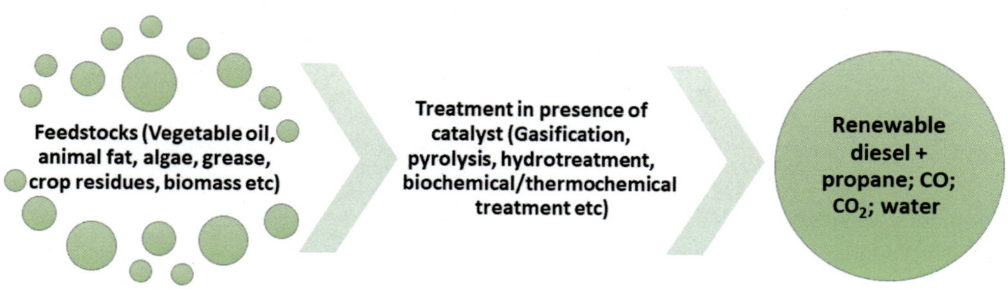

Figure 5.7 Renewable diesel production process.

extreme blending flexibility is responsible for the fuel marketing benefits of renewable diesel in the global market. It enormously impacts GHG emission reduction, mainly NOx and CO_2 emissions (Douvartzides et al., 2019). Currently, renewable diesel has been addressed as a significant alternative for the energy transition. One of the visible differences between renewable diesel and petroleum diesel through burning the fuel has been observed and reported in an article written by Danielle Halprin (2018). They compared the smoke emission from burning renewable diesel and petroleum diesel. Renewable diesel during burning has no observed smoke, while petroleum diesel showed smoke emission that can be distinguished clearly through the results.

5.6 Renewable Fuel Standards and regulation

Renewable Fuel Standard (RFS) program has formed under the Energy Policy Act (EPA) of 2005 and modified by Clean Air Act (CAA). The Energy Independence and Security Act (EISA) 2007 further modified this CAA by expanding the RFS program. The RFS program has been based on national policy to reduce the consumption of petroleum-based transportation fuel with renewable fuel. There are four categories for renewables under RFS (United States Environmental Protection Agency, 2022):

1. Advanced biofuel
2. Biomass-based diesel

Advanced biofuel

- Renewable fuel produced from all kind of renewable biomass (Exception is corn starch)
- Hydrotreating including only co-processing with petroluem

Biomass-based diesel

- Renewable fuel produced from vegetable oils, animal fats, waste cooking oil, lignocellulosic biomass etc.(e.g. biodiesel and renewable diesel)
- Hydrotreating excluding co-processing with petroleum

Cellulosic biofuel

- Renewable fuel produced from lignin, hemicellulose or cellulose
- Any process converting cellulosic biomass

Total renewable fuel

- This includes ethanol produced from corn starch and some other qualifying fuel
- Any process using non-cellulosic portions of food waste and non-cellulosic portions of annual cover crops

Figure 5.8 Categories for renewable fuels under RFS. From United States Environmental Protection Agency. (2022). *EPA report for Renewable Fuel Standard program.* Retrieved April 10, 2022, from https://www.epa.gov/renewable-fuel-standard-program/overview-renewable-fuel-standard.

3. Cellulosic biofuel
4. Total renewable fuel

Fig. 5.8 shows these categories in detail based on processes and feedstocks used for renewable fuel production.

On the other hand, EISA (2007) act showed a considerable increase in the program scope and elaborated some key changes, such as:

- Prolonging requirements of annual volume
- Increasing long-term target up to 36 billion gallons for renewable fuel
- Involving particular types of disclaimer authorities
- Adding clear description for renewable fuel quality regarding GHG emission, renewable biomass, etc.

Volume standards set by EISA for the past 10 years are shown in Table 5.5. The table shows standards for conventional biofuel, advanced biofuel, biomass-based diesel, cellulosic biofuel and total renewable fuel volume. For conventional biofuel, there is no specific requirement for legislative volume. Some biofuels are excluded from the advanced category as they do not qualify for the category due to certain limitations (United States Environmental Protection Agency, 2022).

The CAA provides authority to EPA to regulate advanced, cellulosic and total volumes as a part of the annual statute process. The statute also encloses a general disclaimer authority that sanctions the administrator to relinquish RFS volumes, entirely

Table 5.5 Volumes standards set by EISA.

Year	Conventional biofuel	Advanced biofuel	Biomass-based diesel	Cellulosic biofuel	Total renewable fuel
2022	15.0	21.0	[a]	16.0	36.0
2021	15.0	18.0	[a]	13.5	33.0
2020	15.0	15.0	[a]	10.5	30.0
2019	15.0	13.0	[a]	8.5	28.0
2018	15.0	11.0	[a]	7.0	26.0
2017	15.0	9.0	[a]	5.5	24.0
2016	15.0	7.25	[a]	4.25	22.25
2015	15.0	5.5	[a]	3.0	20.5
2014	14.4	3.75	[a]	1.75	18.15
2013	13.8	2.75	[a]	1.0	16.55
2012	13.2	2.0	1.0	0.5	15.2

[a]Statute sets a minimum requirement of 1 billion gallons, but EPA may increase the requisite.

or partially, and deliberately based on program execution, whether it is responsible for severe economic or environmental harm or less domestic supply.

Different countries have set different essential regulations towards renewable diesel usage. For instance, in 2018 European Commission brought a new ordinance in which the supply of a minimum of 14% renewable energy for rail and road was mandatory from fuel suppliers through 2030. Among that, 3.5% should be from biomass and waste feedstock–derived biofuels. This strategy included RED-II (renewable energy directive-II). In addition, the United States of America has sanctioned policies to support renewable diesel demand and established renewable diesel fuel standards. These specified standards are Renewable Identification Number (RIN) for US Renewable Fuel Standards (RFS) and Low Carbon Fuel Standards (LCFS credits) in California. This program requires blending renewable fuels with diesel through a Renewable Identification Number (RIN) and is a consistent tradable credit system. Renewable diesel meets ASTM D975 specifications for petroleum in the United States, as the level of isomerisation affects fuel quality. So far, up to 70% of isomerisation has been approved.

Furthermore, state-level regulations have been sanctioned to reduce carbon intensity in California from transportation fuel programs called Low Carbon Fuel Standard (LCFS). The target of this program was to reduce fuel carbon intensity by 10% and 30% during 2020 and 2030, respectively. Due to such a supportive and effective platform, renewable diesel supply increased in the United States. The overall supply extended up to 100 million gallons in 2018. Most US biomass-based diesel has been obtained from renewable diesel imports, predominantly from Singapore, since 2015. However, in 2019 this import from Singapore reached 49%, 17,000 barrels/day. Eventually, the overall import for the United States, including biodiesel and renewable diesel, increased by 26%, more than 27,000 barrels/day (US Energy Information Administration, 2020). This example ascertains that fuel standards and regulations are crucial in adapting and commercialising renewable fuels for various sectors.

5.7 Commercial producers of renewable diesel

The commercial segments for renewable diesel production have been installed worldwide. More than 5.5 billion litres of renewable diesel are produced and projected to increase to

13 billion litres by 2024. Several leading companies are producing renewable diesel on a larger scale. Some of them with fuel type, annual production and production process with location are summarised in Table 5.6. The table shows technical details of the process used for renewable diesel production with annual production data. In the past few years, NESTE has produced approximately 50% of renewable diesel and has become the leading renewable diesel producer in the global market.

Besides, so far, worldwide, various renewable diesel plants have been installed. The world is moving towards renewable energy alternatives as an energy transition is necessary for a fossil fuel–free society. Table 5.7 shows the data for renewable diesel plants installed and planned worldwide. Some regions have already reached their target and set expansion goals for the future, presented in Table 5.7. At the same time, some regions are still working to reach the goal.

Table 5.6 Commercial renewable diesel production through bio-feedstock refining (Zhang et al., 2018).

Fuel type	Annual production (million tonnes)	Process	Location	References
Renewable diesel	2.6	NEXBTL process	Neste (Singapore, Finland, Holland)	Neste Oil (2017)
Green diesel	0.5–1	Ecofining process	Eni S.p.A., Italy	Honeywell UOP (2017)
Renewable diesel	0.5	Vegan technology by Axens	Total S.A., France	Kotrba (2015)
Renewable jet fuel	0.5	UOP Renewable Jet Fuel Process	Petrixo Oil & Gas, UAE	Honeywell UOP (2017)
Honeywell green diesel	0.4	Ecofining process	Diamond Green, United States	Diamond Green Diesel (2017)
Renewable diesel	0.2	Bio-Synfining	Renewable Energy Group, United States	Renewable Energy Group report (2017)
Renewable jet fuel	0.15	UOP Renewable Jet Fuel Process	AltAir Paramount, United States	Honeywell UOP (2017)
BioVerno renewable diesel	0.1	HydroFlex hydrotreating process	UPM, Finland	UPM (2015)
Preem evolution diesel	—	HydroFlex hydrotreating process	Preem, Sweden	Preem Evolution Diesel report (2016)
Renewable diesel	—	Hydrotreating	Cetane Energy LLC, United States	Cetane Energy LLC report (2014)

Table 5.7 Worldwide installed and planned renewable diesel plants.

Key player (region)	Target	Status
REG/Phillips66 (Washington)	830,000	Announced (cancelled)
World Energy (California)	125,000	–
Diamond green (Louisiana)	400,000	+275,000 Expansion
Emerland (Louisiana)	365,000	Under construction
REG (Louisiana)	250,000	–
Petrobras (Brazil)	280,000	–
Preem (Sweden)	280,000	Co-processing
UPM (Finland)	130,000	–
NESTE (Finland)	260,000	–
Total (France)	750,000	–
NESTE (Netherlands)	1,000,000	–
CEPSA (Spain)	180,000	Co-processing
REPSOL (Spain)	60,000	Co-processing
ENI Venice (Italy)	360,000	+240,000 Expansion
ENI Gela (Italy)	750,000	–
Petrixo (UAE)	400,000	Under construction
SINOPEC (China)	20,000	–
NESTE (Singapore)	1,000,000	Announced (cancelled)

Source: *From Futurebridge Analysis, EC Reports European Commission, Heuvel, E., et al. (2018).* Building up the future, technology status and reliability of the value chains. *Sub group on advanced biofuels: Sustainable transport forum. Retrieved April 10, 2022, https://data.europa.eu/doi/10.2832/796846 (Futurebridge Analysis, 2018)*

In addition, some other planned renewable diesel production capacities will take over the global market in the future. For instance, project diamond green diesel's extension in Louisiana, Neste's expansion project in Singapore, NEXT renewable fuels Oregon project in the United States and ENI's Venice project expansion in Italy are intended in coming years. As a result, the planned extensions towards renewable diesel production plants have a considerable prospective for the growth of renewable diesel in the global market.

5.8 Worldwide renewable diesel production capacity

Because of the suitable properties of renewable diesel, the worldwide production capacity has been increasing since 2019 and is anticipated to continue this increment until 2022. Fig. 5.9

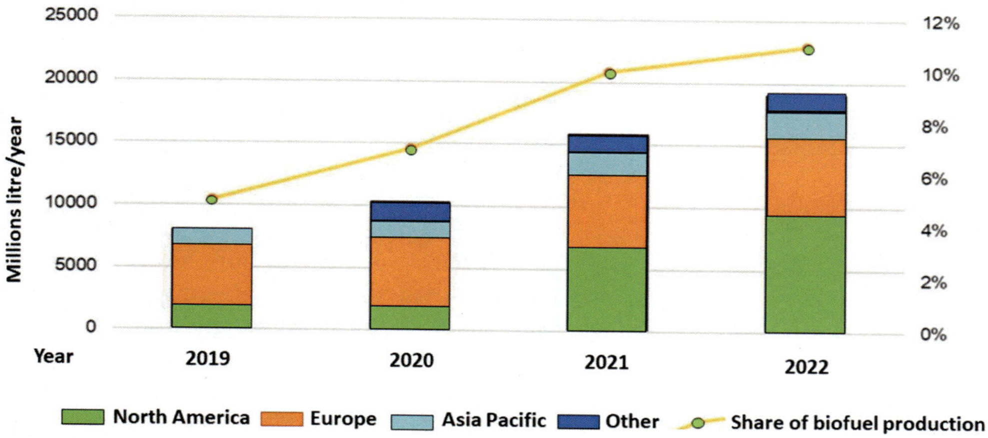

Figure 5.9 Global production capacity for renewable diesel from 2019 to 2022. From International Energy Agency. (2021). *Renewable energy market update: Outlook for 2021 and 2022.* Retrieved April 5, 2022, from https://iea.blob.core.windows.net/assets/18a6041d-bf13-4667-a4c2-8fc008974008/RenewableEnergyMarketUpdate-Outlookfor2021and2022.pdf.

shows the global renewable diesel production capacity trend from 2019 to 2022 (International Energy Agency, 2021).

The major increase in production capacity is mainly for North America, followed by Europe and Asia Pacific. The continuous growth in the production capacity of renewable diesel confirms an optimistic indication for the substitution of petroleum diesel and conventional biodiesel due to its beneficial characteristics.

5.9 Appealing future

The ongoing and growing urbanisation has resulted in a situation where continued search for alternate renewable resources is the only choice to survive for a longer term. With this, climate change due to temperature rise, GHG emission, sea level rise and extreme weather activities are quite disturbing concerns. Numerous technologies for the transportation sector are booming and continues to change to meet the rising demand for clean, reliable and efficient source through the inclusive sector. Hence, globally an energy transition is the most decisive way to accomplish this increasing demand for future transport besides other economic segments. Some developing technologies have the potential to substitute conventional diesel for different applications. Governmental support and encouragement for renewable diesel applications will considerably affect the overall progress and status to fulfil the greatest challenge of accomplishing cleaner fuel for the commercialisation sector. The key approach to future state fuel will be based on zero-emission, advanced energy efficiency, abundant use of renewables and possible hybridisation methods.

The integrated energy transition is a more sustainable and consistent alternative, becoming an expected solution to overcome these challenges. Consequently, efforts through advanced diesel engines and renewable alternatives are imperative to meeting the climate challenge. As high-quality and affordable renewable diesel is already available in the market and verified to be suitable for existing and new vehicles, they also show an extensive reduction in GHG emissions through several sectors that depend extensively on diesel. Commercially produced renewable diesel displays similar engineering standards to conventional petroleum diesel and is applicable for 100% replacement for petroleum fuel.

5.10 Conclusion

Fossil fuels are one of the key energy resources, but due to the increased demand, enormous pressure on available resources exists. It will ultimately increase the economic, environmental and social challenges. Heavy consumption of petroleum causes the depletion of natural fossil fuel reservoirs. Besides, increased dependency on the global market, consequently raising fuel costs and pollution due to GHG gases, suspended particulate matter and organic compounds are the major contributors to global warming, one of the foremost disputes. Hence, substituting fossil fuel is today's necessity, and renewable diesel is the approaching potential candidate to fulfil this necessity. Renewable diesel has several advantages; hence, different countries set various regulations for its use. This motivates the countries to focus on renewable diesel production. To a certain extent, the installed and planned renewable diesel production is increasing, and that makes the baseline for the future of alternative and sustainable fuel. The effort to replace fossil fuel with renewable diesel is a crucial approach to the climatic challenge.

References

Aatola, H., Larmi, M., Sarjovaara, T., Mikkonen, S. (2008). Hydrotreated vegetable oil (HVO) as a renewable diesel fuel: trade-off between NOx, particulate emission, and fuel consumption of a heavy duty engine. *SAE Technical Paper*, p.12; 2008-01-2500. Available from https://doi.org/10.4271/2008-01-2500

Agwu, O., Valera-Medina, A., Katrašnik, T., & Seljak, T. (2021). Flame characteristics of glycerol/methanol blends in a swirl-stabilised gas turbine burner. *Fuel, 290*, 119968. Available from https://doi.org/10.1016/j.fuel.2020.119968.

Blog from Honeywell UOP 2017. Honeywell Green Diesel™ <https://www.uop.com/processing-solutions/renewables/green-diesel/>. (Accessed 10.04.2022).

Cetane Energy LLC report 2014 <http://www.cetaneenergy.com/>. (Accessed 10.04.2022).

Climate Watch, based on raw data from IEA 2021, *GHG Emissions from Fuel Combustion*, 2022. www.iea.org/statistics; modified by WRI <https://www.wri.org/insights/4-charts-explain-greenhouse-gas-emissions-countries-and-sectors> (Accessed 3.04.2022).

Chia, S. R., Nomanbhay, S., Ong, M. Y., Shamsuddin, A. H. B., Chew, K. W., & Show, P. L. (2022). Renewable diesel as fossil fuel substitution in Malaysia: A review. *Fuel, 314*, 123137. Available from https://doi.org/10.1016/j.fuel.2022.123137.

Data from IEA. (2020) modified by WRI, <www.iea.org/statistics>. (Accessed 3.04.2022).

Douvartzides, S. L., Charisiou, N. D., Papageridis, K. N., & Goula, M. A. (2019). Green diesel: Biomass feedstocks, production technologies, catalytic research, fuel properties and performance in compression ignition internal combustion engines. *Energies, 12*(5), 809. Available from https://doi.org/10.3390/en12050809.

Fossil fuels and climate change: the facts. (2022). <https://www.clientearth.org/> (Accessed 3.04.2022).

Friedlingstein, et al. 2021, *Global carbon budget 2021*, <https://doi.org/10.5194/essd-2021-386> (Accessed 3.04.2022).

Gebremariam, S. N., & Marchetti, J. M. (2018). Economics of biodiesel production: Review. *Energy Convers Manage, 168*, 74. Available from https://doi.org/10.1016/j.enconman.2018.05.002.

Heuvel, E., et al. (2018). *Futurebridge analysis*. EC Reports European Commission, Building up the future, technology status and reliability of the value chains: sub group on advanced biofuels: sustainable transport forum, Publications Office. <https://data.europa.eu/doi/10.2832/796846> (Accessed 10.04.2022).

IEA. (2021). *Global Energy Review 2021*, IEA, Paris. <https://www.iea.org/reports/global-energy-review-2021/co2-emissions>. (Accessed 03.04.2022).

Integrated Flow Solutions. (2020). *Biodiesel guide – sources, production, uses, & regulations*. <https://ifsolutions.com/how-is-biodieselmade-sources-uses/>. (Accessed 03.04.2022).

International Energy Agency. (2021). *Renewable energy market update: outlook for 2021 and 2022*. <https://iea.blob.core.windows.net/assets/18a6041d-bf13-4667-a4c2-8fc008974008/RenewableEnergyMarketUpdate-Outlookfor2021and2022.pdf>. (Accessed 05.04.2022).

International Energy Agency. (2020). *IEA Renewables 2020*. <https://www.iea.org/reports/renewables-2020/transport-biofuels> (Accessed 05.04.2022).

Karavalakis, G., Hilari, D., Givalou, L., Karonis, D., & Stournas, S. (2011). Storage stability and ageing effect of biodiesel blends treated with different antioxidants. *Energy, 36*(1), 369. Available from https://doi.org/10.1016/j.energy.2010.10.029.

Magazine Total to convert oil refinery to renewable diesel production by Kotrba R. 2015 <http://www.biodieselmagazine.com/articles/355201/total-to-convert-oil-refinery-to-renewable-diesel-production> (Accessed 10.04.2022).

Michigan farm news report. (2022). <https://www.michiganfarmnews.com/biodiesel-vs-renewable-diesel-and-the-growing-demand-for-both?__cf_chl_tk = 7wKCSbYJ6Wu998hfMEJNH_u5CcMF6hSEnpvJ0c.btQ0-1649738656-0-gaNycGzNCNE>. (Accessed 05.04.2022).

Neste Oil. (2017). *Neste's renewable diesel celebrates 10 years of reducing emissions.* <https://www.neste.com/en/nestes-renewable-diesel-celebrates-10-years-reducing-emissions> (Accessed 10.04.2022).

Nomanbhay, S., Hussein, R., & Ong, M. Y. (2018). Sustainability of biodiesel production in Malaysia by production of bio-oil from crude glycerol using microwave pyrolysis: a review. *Green Chem Lett Rev, 11*(2), 135. Available from https://doi.org/10.1080/17518253.2018.1444795.

Preem Evolution Diesel report. Preem 2016 <http://preem.se/en/in-english/investors/annual-overview-2016/sustainable-development/preem-evolution-fuels/preem-evolution-diesel/> (Accessed 10.04.2022).

Rapier, R. (2008). Renewable Diesel. In D. Pimentel (Ed.), *Biofuels, Solar and Wind as Renewable Energy Systems.* Dordrecht: Springer. Available from https://doi.org/10.1007/978-1-4020-8654-0_7.

Renewable Energy Group report 2017. *Bio-Synfining* <http://www.regi.com/technologies/bio-synfining>. (Accessed 10.04.2022).

Report by Danielle Halprin. (2018). *CCJPA Climate Corp Fellow 2017-18 for Strategic Energy Innovations.* <https://www.capitolcorridor.org/blogs/get_on_board/ccjpa-tests-renewable-diesel/> (Accessed 05.04.2022).

Report from Diamond Green Diesel 2017 <https://www.diamondgreendiesel.com/> (Accessed 10.04.2022).

Report from UPM. (2015). *UPM Lappeenranta Biorefinery is in commercial production.* <http://www.upm.com/About-us/Newsroom/Releases/Pages/UPM-Lappeenranta-Biorefinery-is-in-commercial-production-001-Mon-12-Jan-2015-11-30.aspx> (Accessed 10.04.2022).

Statistical Review of World Energy 2021 | 70th edition, BP energy outlook <https://www.bp.com/en/global/corporate/energy-economics/statistical-review-of-world-energy.html> (Accessed 05.04.2022).

The Intergovernmental Panel on Climate Change (IPCC) report. (2022). <https://www.ipcc.ch/> (Accessed 03.04.2022).

United States Environmental Protection Agency. (2022). *EPA report for Renewable fuel standard program.* <https://www.epa.gov/renewable-fuel-standard-program/overview-renewable-fuel-standard>. (Accessed 10.04.2022).

U. S. Department of Energy. (2021). *Ethanol fuel basics.* <https://afdc.energy.gov/fuels/ethanol_fuel_basics.html> (Accessed 05.04.2022).

U. S. Energy Information Administration. (2020). *EIA report 2020* <https://www.eia.gov/todayinenergy/detail.php?id = 43935> (Accessed 10.04.2022).

U. S. Energy Information Administration. (2021a). *Use of energy explained.* <https://www.eia.gov/energyexplained/use-of-energy/>. (Accessed 03.04.2022).

U. S. Energy Information Administration. (2021b). *Use of energy, transportation explained.* <https://www.eia.gov/energyexplained/use-of-energy/transportation.php> (Accessed 05.04.2022).

Zhang, B., Wu, J., Yang, C., et al. (2018). Recent developments in commercial processes for refining bio-feedstocks to renewable diesel. *Bioenerg Res, 11*, 689. Available from https://doi.org/10.1007/s12155-018-9927-y.

6

Renewable diesel and biodiesel: a comparative analysis

Arpita Das and Samuel Lalthazuala Rokhum
Department of Chemistry, National Institute of Technology Silchar, Silchar, Assam, India

6.1 Introduction

The search for alternate energy sources has intensified due to growing concerns. These worries include the availability of feedstock in relation to supply security and the utilisation of domestic energy sources, volatility of cost, the ongoing depletion of non-renewable petroleum reserves, and greenhouse gas emissions. It is beyond the scope of this chapter to list all of the legal, regulatory, and incentive measures implemented to address these concerns. The importance of fuels generated from biological sources, including lipid substances like fats and oils, has grown. Fuels produced through various production methods employing fats and oils as feedstocks have varying compositions and physical characteristics. Biodiesel is the most well-known of these fuels (Agrawal, Belkhode, & Rokhum, 2023; Changmai, Vanlalveni, Ingle, Bhagat, & Rokhum, 2020), formed by transesterifying an oil or fat with alcohol from vegetable or animal sources which is also described as a mixture of mono-alkyl esters. Its high viscosity (often between 28 and 40 mm^2/s), which causes operational issues in diesel engines such as deposit production and injector coking upon injection into the combustion chamber due to poorer atomisation, is the main deterrent against using plain vegetable oil as fuel. Oil viscosity lowers during transesterification to a range (typically 4–5 mm^2/s) closer to petrodiesel. But in current years, curiosity has grown about a fuel called "sustainable diesel" and whose makeup is similar to that of diesel fuel made from petroleum (petrodiesel). To manufacture fuels that resemble petrodiesel, several techniques (cracking or pyrolysis, HDO) can be utilised (Huber & Corma, 2007). A flowchart for manufacturing fuels from feedstocks containing triacylglycerol (TAG) to engine combustion is shown in Fig. 6.1.

Renewable Diesel. DOI: https://doi.org/10.1016/B978-0-323-91153-5.00002-9

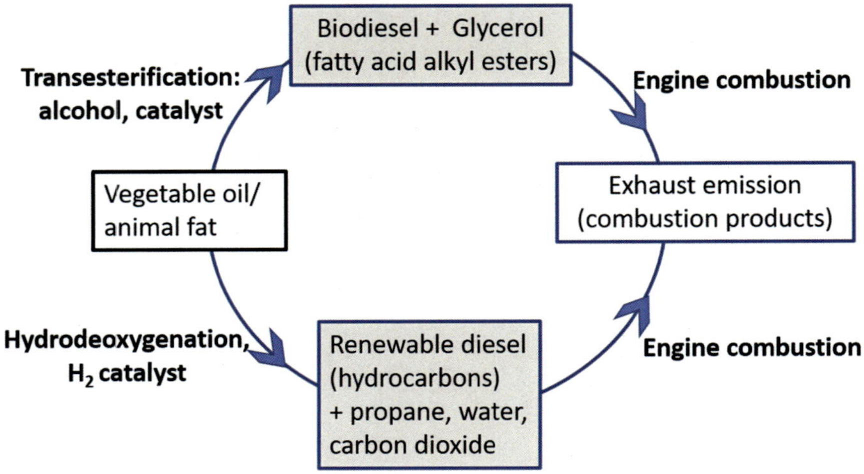

Figure 6.1 A flowchart showing the conversion of lipid ingredients (biodiesel and renewable diesel by hydrodeoxygenation) into engine combustion products. Adapted from Knothe, G. (2010). Biodiesel and renewable diesel: A comparison. *Progress in Energy and Combustion Science, 36*, 364–373. https://doi.org/10.1016/j.pecs.2009.11.004.

6.1.1 Terminology and a brief description of different types of fuel

A literature survey reveals that, occasionally, distinct terminologies are used to represent the same type of fuel obtained from a feedstock containing triacylglycerol. Conversely, that single keyword is used to describe a variety of fuels eventually. Therefore the terminology is crucial to characterise the areas of discussion and is briefly addressed here, analogous to the issue of terminology in chemistry (Knothe, 2010).

1. *Petrodiesel*—The diesel fuel made from petroleum must first be defined since it serves as the benchmark against which all other fuels are measured. It can be referred to as "petrodiesel." Although it is typically referred to as "diesel" or "diesel fuel" in practically all literature, the prefix "petro-" indicates its origin in comparison to the current alternatives. In the United States, ASTM D975 (Semorile, Alviso, & Romano, 2023) and in Europe, EN 590 (European Standard, 2013) are the petrodiesel standards that are frequently used.
2. *Biodiesel*—The term "biodiesel" refers to a fuel that is composed of mono-alkyl esters of long-chain fatty acids derived from vegetable oils or animal fats, designated as "B100" in the biodiesel standard ASTM D6751 (Regulation, 2010); the European biodiesel standard EN 14214 (European Committee for Standardization, 2014) refers to fatty acid methyl esters

(FAME) as a fuel. This fuel is derived from oil or fat via a transesterification process with glycerol as a byproduct. One could argue that only about 95% of the carbon atoms in biodiesel are "bio" when it is created with methanol as the alcohol component. The reason behind it is that although methanol can be obtained from renewable resources, it is most frequently produced using non-renewable natural gas. In more recent times, the word "biodiesel" has occasionally been combined with "second-generation biodiesel," typically in reference to biodiesel made from "alternative" feedstocks like inedible/waste oils or algae.

3. *Renewable diesel (RD)*—In addition to the term "second-generation biodiesel" stated ealier, the names "green diesel" and "RD" have also been used to refer to biofuels that resemble petrodiesel. Because it indicates that the final fuel is "greener" than petrodiesel, the term "green diesel" is vague. The phrase "RD" seems to make no other inferences about the fuel's characteristics other than that it comes from a renewable resource. As a result, it seems to be the appropriate term to describe fuels that resemble petrodiesel but originate from biological sources. It should be noted that some technical literature refers to RD as "hydrotreated vegetable oil" (HVO).

This chapter addresses several problems associated with a comparison of biodiesel and RD. These issues comprise an introduction to both types of diesel, their production pathways, fuel properties, combustion characteristics, challenges, and opportunities.

6.1.1.1 Biodiesel

Rising energy demands have worsened climate change and global warming in addition to fully depleting non-renewable petroleum supplies, which are detrimental to people in the 21st century (Chuah, Klemeš, Yusup, Bokhari, & Akbar, 2017). As a result, the hunt for alternative renewable energy has been sparked by worries about the depletion of fossil fuels and environmental degradation. Researchers have focused on biodiesel as one of the contenders because it is a promising alternative fuel (Kumawat & Rokhum, 2022; Pan et al., 2022). Biodiesel is a green sustainable, and environmentally friendly biofuel that could replace petroleum-based diesel as a source of energy (Pölczmann, Tóth, Beck, & Hancsók, 2016). As stated in the biodiesel standard ASTM D6751 (Alleman, Christensen, & Moser, 2019), the term "biodiesel" refers to a fuel of long-chain fatty acids comprised of mono-alkyl esters derived from animal fats or vegetable oils, designated as "B100," with the European biodiesel standard EN 14214 (European Committee for Standardization, 2014) designating FAME as the fuel. Glycerol is a co-product of the transesterification process used to create this fuel from an oil or fat. This glycerol is further transformed into value-added derivatives such as solketal (Ao & Rokhum, 2022; Saikia et al., 2022) and glycerol carbonate (Changmai, Laskar, & Rokhum, 2019; Das, Shi, Halder, & Lalthazuala, 2022) as a fuel additive.

Most biomass used to make biodiesel comes from renewable sources, including edible and inedible oils, animal fats, used cooking oils, oleaginous micro-organisms, and fats. They can be divided into four groups based on the source of the feedstock. We can better comprehend the procedure for producing many generations of biodiesel by understanding the precise purity and composition of each feedstock, as shown in Fig. 6.2. Today, a number of nations have laws and policies that encourage the use of alternative fuels, including biodiesel, through both financial incentives and legally enforceable requirements. Ethanol and biodiesel made up 3% of the world's transportation energy in 2020, an increase of 5% from 2019 (Ogunkunle & Ahmed, 2019). Despite increasing interest in the electric transportation system, biodiesel of 336.5 million barrels was generated globally

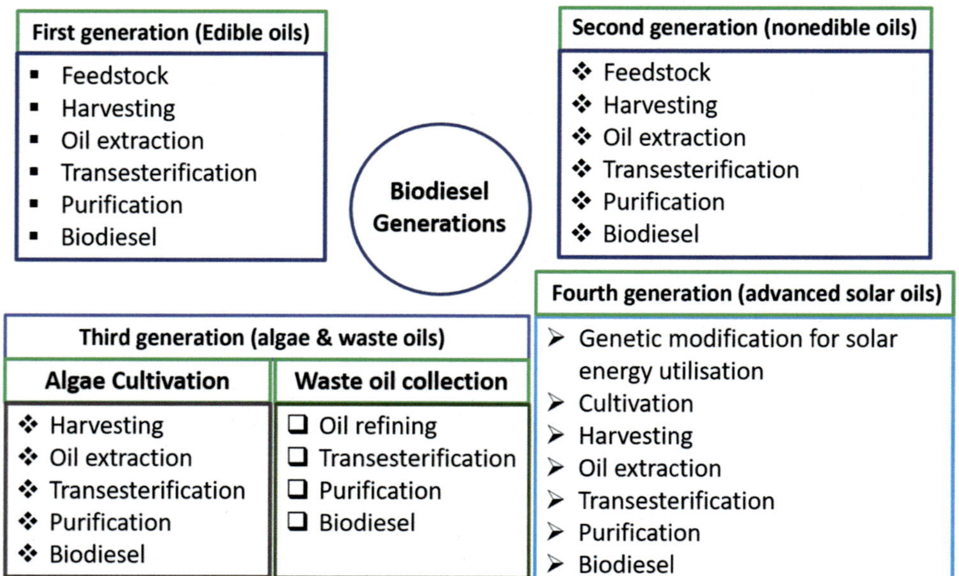

Figure 6.2 Different generations of biodiesel depending on feedstock source. Adapted from Bashir, M. A., Wu, S., Zhu, J., Krosuri, A., Khan, M. U., & Ndeddy Aka, R. J. (2022). Recent development of advanced processing technologies for biodiesel production: A critical review. *Fuel Processing Technology, 227*, 107120. https://doi.org/10.1016/j.fuproc.2021.107120 (Bashir et al., 2022).

in 2017 (REN21, 2017), accounting for more than 90% of all sustainable transport fuels. The life cycle of sustainable biodiesel manufacturing is depicted in Fig. 6.3.

The physicochemical processes used to make biodiesel include transesterification, esterification, pyrolysis, and supercritical fluids. The most popular technique of producing biodiesel is transesterification, which involves reacting triglycerides (TGs) and alcohol with catalysts to form fatty acid alkyl ester and glycerol. Depending on the process of treatment, a variety of raw materials, including animal fats, vegetable oils, waste cooking oils, microbial oils, and algal oils, can be utilised as feedstock in the manufacturing of biodiesel (Baskar & Aiswarya, 2016; Bet-Moushoul et al., 2016; Chhandama, Satyan, Changmai, Vanlalveni, & Rokhum, 2021; Ibrahim et al., 2022). Additionally, an exceptional feedstock for biodiesel synthesis is burnt biodiesel obtained from vegetable oil, demonstrating how using biodiesel may significantly cut greenhouse gas emissions while being economically advantageous (Ogunkunle & Ahmed, 2019). The main reason why biodiesel produced today cannot compete economically with petroleum-based diesel in the transportation market is the efficiency of production that has largely reached its ceiling with the

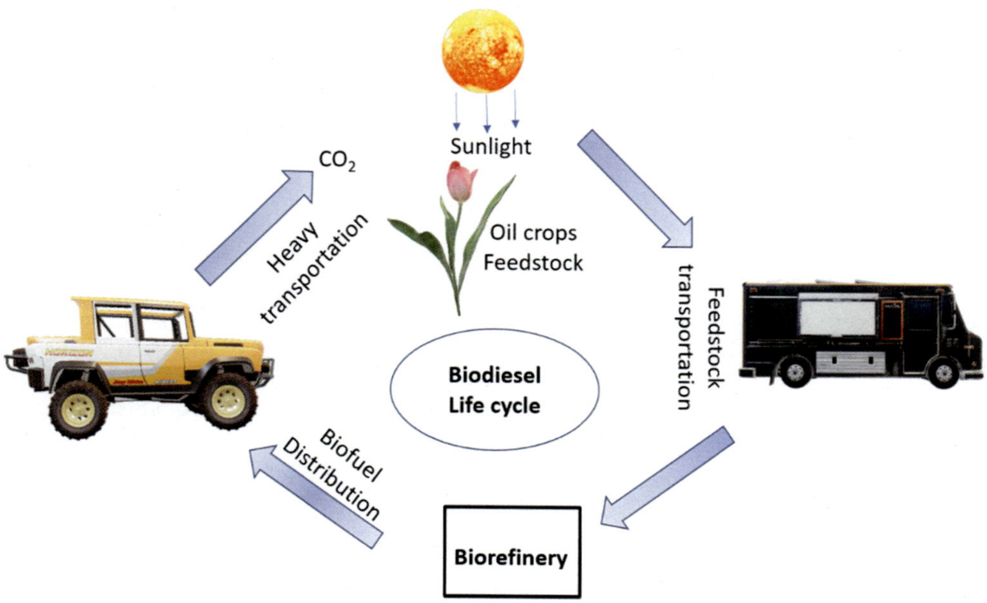

Figure 6.3 Life cycle of sustainable biodiesel production. Adapted from Bashir, M. A., Wu, S., Zhu, J., Krosuri, A., Khan, M. U., & Ndeddy Aka, R. J. (2022). Recent development of advanced processing technologies for biodiesel production: A critical review. *Fuel Processing Technology, 227*, 107120. https://doi.org/10.1016/j.fuproc.2021.107120.

rise in production costs and little room for further improvement. However, traditional production technologies are a major pathway for biodiesel production, viz. transesterification (Badday, Abdullah, Lee, & Khayoon, 2012). As a result, in recent years, there has been a greater demand for the manufacture of biodiesel using innovative, clean, and enhanced methods. The reaction times of these new, environmentally friendly technologies must be sped up while still producing biodiesel of high calibre. In order to do this, a variety of techniques have lately been investigated, including microwave (Changmai et al., 2019; Das, Shi, Halder, & Lalthazuala Rokhum, 2022; Motasemi & Ani, 2012; Soltani, Rashid, Yunus, & Taufiq-Yap, 2015), membrane (Shuit, Ong, Lee, Subhash, & Tan, 2012), ultrasonic (Badday et al., 2012; Luo, Fang, & Smith, 2014; Rajkumari et al., 2021), hydrodynamic cavitation (Chuah, Yusup, Abd Aziz, Bokhari, & Abdullah, 2016), plasma discharge (Troter, Todorović, Dokić-Stojanović, Stamenković, & Veljković, 2016), reactive distillation, cosolvent, rotatory, supercritical methods (Tan & Lee, 2011), and plug flow reactors (Buchori, Istadi, & Purwanto, 2016) for the generation of biodiesel (Laskar et al., 2018).

6.1.1.2 Renewable diesel

The term "RD" refers to "diesel-like alkanes" made from vegetable oil. It is frequently mentioned as second-generation biodiesel or HVO, also called green diesel, and shares long-chain alkanes with mineral diesel in terms of composition. Lipids and biodiesel can be hydrotreated to create RD. Compared to biodiesel, RD has improved cold-flow characteristics and is a relatively clean fuel. For diesel-like alkanes made from renewable sources such as animal fat, vegetable oil, single cell oil (SCO) from oleaginous microorganisms (including some fungi, algae, yeast, and bacteria) or even biodiesel, by HDO, Knothe claims that the term RD appears to be the most appropriate (Knothe, 2010). HDO is one of the five technological methods for reducing the viscosity of TAGs. Pyrolysis, microemulsion, mixing, and transesterification are the other four. Using a catalyst, HDO is carried out at high temperatures and pressures when H_2 is present. In the process, hydrogen is substituted for moieties of ester (-OR) and double bonds, resulting in end products that resemble alkanes. As HDO catalysts, $NiMo/Al_2O_3$, $CoMo/Al_2O_3$, $Rh\text{-}Al_2O_3$, CoMo/Si, Pd/C, Pd/SiO_2, Pt/C, CoMo/C, and Ni/SiO_2 have all been employed. RD has greater cold flow characteristics and oxidative stability than biodiesel. Thus it can be produced and delivered utilising the current petroleum distribution and refining infrastructure (Srifa, Faungnawakij, Itthibenchapong, & Assabumrungrat, 2015). Despite these benefits, additional study is needed to determine RD's economic, energy, and environmental performance using life cycle assessment-based techniques. Biodiesel performs better regarding carbon monoxide (CO) emissions and particulate matter, while RD is cleaner regarding NOx and hydrocarbon emissions than biodiesel (Arun, Sharma, & Dalai, 2015).

Over the past 10 years, advanced methods have been developed and made available for turning bio-feedstocks into sustainable diesel. RD production capacity in the world is getting close to 5.5 million tonnes yearly. To eliminate impurities, pretreatment of the renewable feedstock, hydroprocessing and isomerization to create hydrocarbons, and distillation to create a fuel suitable for usage as jet or diesel fuel are all parts of the refining process. Chemically similar to petroleum diesel, RD is a blend of paraffin, straight-chain, and branched with typical carbon values of $C_{15}-C_{18}$ (Neste Corporation, 2015). As a result, engines that use conventional diesel fuel can run on RD as a drop-in fuel (Wicaksana, 2016). As an alternative, fatty acid methyl/ethyl esters from vegetable oils and animal fats can be converted into biodiesel (Chen, 2017; Kim, Hanifzadeh, & Kumar, 2018).

In other words, a synthetic diesel fuel made from the hydrogenation of fatty acids containing no oxygen and aromatic chemicals is renewable. This diesel will be viewed as an alternative to diesel made from a renewable feedstock that is not an ester to maintain some consistency. Chemically, it entails the conversion of TG-rich plant and animal fats into the equivalent alkanes through direct catalytic hydrogenation. Components of propane and gasoline are by-products of this hydrogenation. Any animal fat or vegetable oil can be used as an input to the process to create an aliphatic product with properties resembling Fischer–Tropsch (FT) synthetic fuel. RD is often put through pyrolysis-rapid thermal processing or thermal hydrotreating, although variations may vary depending on the firm. The end product is a clear, colourless hydrocarbon fuel with good cetane numbers and superior combustion qualities to even petroleum-based diesel, in contrast to the yellow transesterified biodiesel. When compared to ultra-low sulphur diesel (ULSD) that has received California Air Resources Board (CARB) certification, 100% Neste Renewable Diesel (formerly NExBTL) was found to have NOx emissions that were reduced by 5%–18% and emissions that were reduced by 5%–28% (Hajbabaei et al., 2012).

6.2 Production pathways

A co-reagent and catalysts are needed to produce the required product from the biodiesel and RD feedstock. The co-reagent in biodiesel is often alcohol (methanol to produce methyl esters), whereas hydrogen is used in RD for double bond saturation and oxygen substitution. The catalyst often used for biodiesel is sodium alkylate (methylate in the case of methanol) or sodium or potassium hydroxide (alkylates are preferred to hydroxides). In contrast, different catalysts are used for RD based on the precise type or desired products. The method known as HDO, however, has likely received the most attention for RD.

6.2.1 Biodiesel processing methods

6.2.1.1 Physical methods

In order to prepare raw feedstock for fuel alternatives in diesel engines, there are four main conventional methods. (1) Blended oils direct employment, (2) micro-emulsion of oils, (3) pyrolysis or thermal cracking of oils, and (4) transesterification. To make oil appropriate for use as biodiesel, all of these production pathways lower oil viscosity against oxidation, increase oil stability, and boost volatility (Sundus, Fazal, & Masjuki, 2017). Alcoholysis is currently the popular method for biodiesel production (Pathak, Das, Rajkumari, & Rokhum, 2018). Other methods result in the lowering of TG viscosity (Rezania et al., 2019). The following methods used to treat the biodiesel are:

6.2.1.1.1 Dilution or direct use of blended oils

In 1893, Rudolf Diesel used straight peanut oil as diesel fuel in a compression engine (Bart, Palmeri, & Cavallaro, 2010). Vegetable oils have undergone various physical and chemical alterations in order to increase their volatility and viscosity for use as diesel fuel because natural oil does not alter its properties. Following issues with using direct vegetable oil in diesel engines, researchers investigated the idea of combining vegetable oil and diesel at various ratios as engine fuel. This was the trend from the late 1920s until 1982 when the first international conference was held on the impact and financial aspects of using bended oil (Ma & Hanna, 1999). Vegetable oil can be utilised directly (or mixed with diesel) at low manufacturing costs, but the mixture needs extensive pre-treatment in order to function properly in diesel engines. It is not feasible to utilise the highly viscous solution directly in engines since it solidifies in cold conditions.

Additionally, the unsaturated hydrocarbons' high-free fatty acid (FFA) content causes more oxidation, which increases engine wear and the accompanying maintenance costs. In order to maintain the proper viscosity, vegetable oil is diluted with petroleum diesel (Muniyappa, Brammer, & Noureddini, 1996). This is because crude vegetable oil has a viscosity 10–17 times greater than petroleum diesel (Gulum, Kutlu Onay, & Bilgin, 2018). Diesel is mixed with vegetable oil to operate engines. To maintain entire power without variation or engine adjustment in pre-combustion chamber engines in 1980, Caterpillar Brazil employed 10-wt.% vegetable oil. Diesel had not yet succeeded in replacing 100-wt.% vegetable oil, but a blend of 20-wt.% vegetable oil and 80-wt.% diesel fuel had. Diesel made up 75% of the fuel, while 25% of it was sunflower oil. Because of the low viscosity, engine efficiency is increased while the amount of fuel is decreased.

6.2.1.1.2 Micro-emulsion

It involves mixing the oil feedstock with the appropriate solvents (Ndiaye, Arhaliass, Legrand, Roelens, & Kerihuel, 2020). Methanol, ethanol, and 1-butanol are the most frequently employed solvents in present-day research (Li et al., 2018). Amphiphilic substances, such as phospholipids, sorbitan monooleate, *n*-butanol, 2-octanol, and carboxylate surfactants, are used to stabilise the dilution and enhance the solubility of alcohols because they have a restricted ability to dissolve in water (Kalsoom, Zerey-belaskri, Nadeem, & Shehzad, 2017). When compared to neat vegetable oil or neat diesel, fuel emulsion emits less nitric oxide and CO (Siva, Munuswamy, & Devarajan, 2019).

A microemulsion is an optically isotropic fluid microstructure dispersion that is described as a colloidal equilibrium, with measurements typically ranging from 1 to 150 and formed spontaneously by two typically immiscible liquids as well as several ionic amphiphiles (Likozar & Levec, 2014). They can improve spray qualities by vapourising all of the micelle's low-boiling ingredients. The engine's performance was unaffected by the micro-emulsion of 25 and 53 wt.% sunflower oil in diesel. The soyabean oil, cetane, 2-octanol, and methanol microemulsion, which were made in the ratio of 52.7:1.0:33.3:13.3, likewise passed the 200 h-engine manufacturer's association (EMA) test.

One physical technique to use vegetable oil in a diesel engine without chemical modification is microemulsion (Ma & Hanna, 1999). Vegetable oil, with or without diesel fuel, is mixed with esters and dispersants (solvents) to create microemulsions, which are transparent and thermodynamically stable mixtures of oil and surfactants (Yusuf, Kamarudin, & Yaakub, 2011). Due to the higher alcohol content, the calorific value of microemulsions is lower than diesel. And also, microemulsions can cool the combustion chamber due to the higher latent heat property of these alcohols. Methanol and vegetable oil microemulsions perform nearly as well as diesel fuel. Previous studies examined the influence of co-surfactants and the impact of catalysts in water oil microemulsions made from various refined and high-FFA oils (de Jesus, Silva, & Vale, 2008; Wellert et al., 2008). In a few thermochemical liquefication research, sewage sludge was combined with various co-surfactants to create diesel fuel via microemulsion, improving the fuel's use and enhancing its physicochemical qualities. However, additional study is required to enhance this technique for use on a global scale because of the substantial carbon waste and extremely low efficiency (Ding et al., 2017; Leng, Han, Yuan, Li, & Zhou, 2018).

6.2.1.2 Chemical methods

6.2.1.2.1 Thermal cracking (pyrolysis)

Pyrolysis uses heat or catalysts to change one material into another when there is not any oxygen or air present (Higman, Higman, Chortyk, & Schmeltz, 1973). In comparison to previous cracking techniques, the procedure is quick, clean, and effective (Schwab, Dykstra, Selke, Sorenson, & Pryde, 1988). In Fig. 6.4, the thermal degradation reaction is depicted. New technologies are currently being employed to produce biodiesel because the pyrolysis method is no longer effective. An alternate way of producing biodiesel is called pyrolysis, which is the heat breakdown of big organic molecules into smaller molecules in the absence of oxygen and air (Wang et al., 2020; Xu et al., 2020). There are catalytic and noncatalytic processes in thermal cracking. Two common catalysts utilised in the conversion process

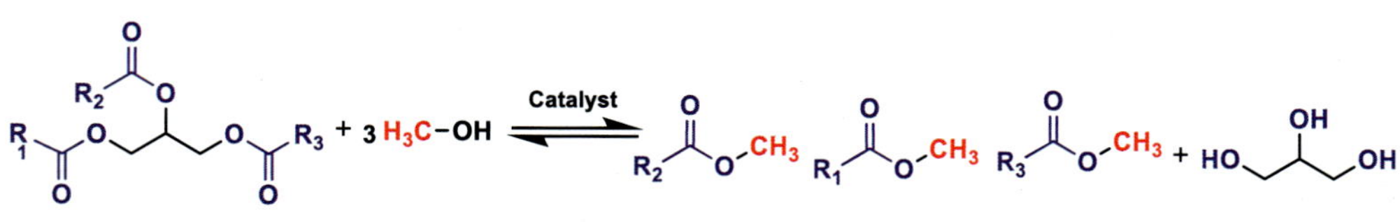

Figure 6.4 Biodiesel synthesis via transesterification reaction and glycerol as a byproduct. Adapted from Bashir, M. A., Wu, S., Zhu, J., Krosuri, A., Khan, M. U., & Ndeddy Aka, R. J. (2022). Recent development of advanced processing technologies for biodiesel production: A critical review. *Fuel Processing Technology, 227*, 107120. https://doi.org/10.1016/j.fuproc.2021.107120.

are zeolite and silica-alumina (Tan, Abdullah, & Hameed, 2018). While the oil is being thermally cracked, oxygen must be removed in order to finish the conversion of the oil into biodiesel. There are not many environmental advantages, and certain low-value materials are produced (Nigam & Singh, 2011).

By applying heat or pyrolysis, which is the process of changing one organic molecule into another, natural oils can be transformed into various esters that can be used as diesel fuel. Vegetable oils, animal fats, natural fatty acids, and FAME are examples of materials that are thermally heated. TG pyrolysis has been the subject of extensive research to create appropriate fuel for diesel engines (Canakci, Ozsezen, Arcaklioglu, & Erdil, 2009). Despite the versatility of the pyrolysis technique for various oils, it has not been properly assessed because of the expensive and complicated equipment. Additionally, the reaction conditions are highly severe, leading to the synthesis of short-chain molecules more similar to gasoline than diesel, and the biofuels generated lack the necessary fuel characteristics. Pyrolysis, a method for turning used cooking oil into biofuel, produces by-products that can be consumed as fertilisers and have a conversion efficiency of about 80% (Aida, Kaouther, Withek, Slim, & Aymen, 2018; Wan Mahari et al., 2018).

6.2.1.2.2 Transesterification

The most popular chemical process for turning natural fats and oils into diesel is transesterification, which involves a stoichiometric reaction between three alcohol molecules and one molecule of TGs (Fig. 6.5). Alkyl esters (biodiesel), which are produced as a byproduct of this process and glycerol as a byproduct, are carried out in the presence of a catalyst (Mustranta, Forssell, & Poutanen, 1993; Najafi et al., 2019; Trejo-Zárraga, Hernández-Loyo, Chavarría-Hernández, & Sotelo-Boyás, 2018; Van Gerpen, Shanks, Pruszko, Clements, & Knothe, 2004). Investigation into the reaction kinetics of TG transesterification revealed that while the total conversion was unaffected by temperature, the transesterification process' speed did. The rate of transesterification was higher at higher temperatures as the time required for mass transfer was reduced, and the overall reaction kinetics were dependent on the individual constants of the TG conversion rate to diglycerides, monoglycerides, and alcohol esters (Darnoko & Cheryan, 2000).

Alcoholysis is the process by which alcohol displaces an ester in the production of biodiesel, which is typically created through transesterification to produce esters and the by-product

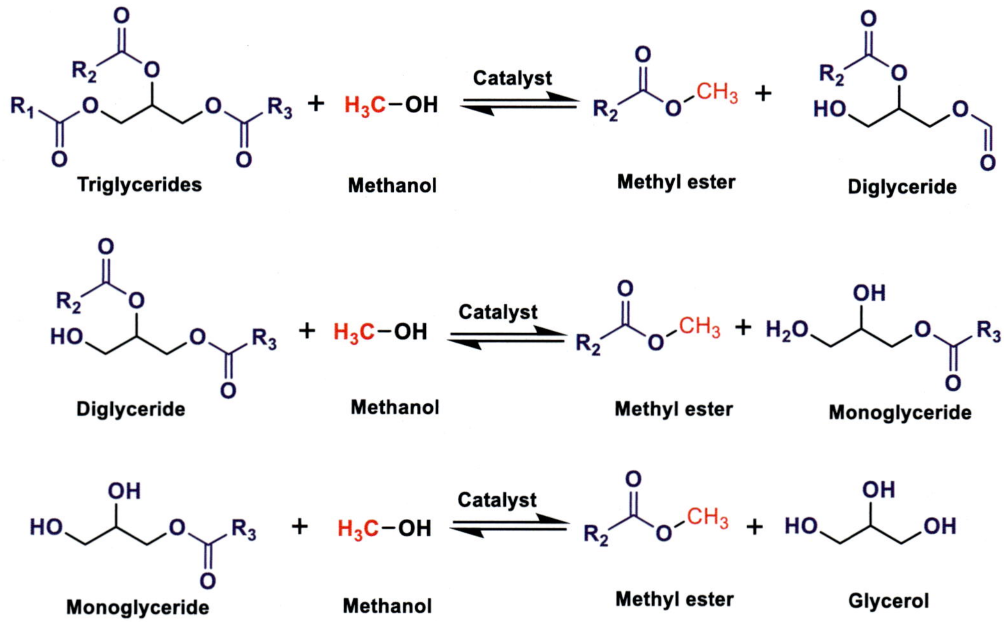

Figure 6.5 Transesterification of mono-, di-, and triglycerides. Adapted from Bashir, M. A., Wu, S., Zhu, J., Krosuri, A., Khan, M. U., & Ndeddy Aka, R. J. (2022). Recent development of advanced processing technologies for biodiesel production: A critical review. *Fuel Processing Technology, 227*, 107120. https://doi.org/10.1016/j.fuproc.2021.107120.

glycerol (alcohol-cleavage). Common alcohols utilised in this procedure include methanol, ethanol, propanol, butanol, and amyl alcohol (Navas, Lick, Bolla, Casella, & Ruggera, 2018). With a lipid or oil-to-alcohol stoichiometry of 1:3, the reaction is reversible, and extra alcohol causes the equilibrium to shift in favour of the forward direction, as shown in the equation below. This ratio is often raised due to the product's higher yield.

It involves three reversible reactions that happen in a row. TGs are transformed into diglycerides in the first stage, then into monoglycerides, and finally into glycerol in the final step (Algoufi, Akpan, Kabir, Asif, & Hameed, 2017). A methyl ester molecule, a molecular type of biodiesel, is produced in each step.

Transesterification is the most advantageous method for synthesising biodiesel due to its efficient viscosity reduction (Silitonga et al., 2020). Through transesterification, the oil feedstock and alcohols are catalysed to produce glycerol and biodiesel (fatty acid alkyl esters) as the final product (Al-Saadi, Eze, & Harvey, 2018). The excess alcohol is recovered and recycled back into the first step of the transesterification process following the separation of the byproduct and the final product.

Overall, it is a low-temperature, ecologically benign process that can convert a range of oil feedstock. Temperature, time, pressure, the amount of alcohol in the feedstock oil, concentration, kind of catalyst, mixing intensity, and feedstock oil are specific process variables that have an impact on the transesterification reaction.

Utilising primary or secondary monohydric aliphatic alcohols, the transesterification reaction can happen with or without a catalyst (Mueanmas, Nikhom, Petchkaew, Iewkittayakorn, & Prasertsit, 2019). It should be noted that transesterification processes can happen without catalysts. In addition to pre- and post-treatment steps, they necessitate high temperatures, pressures, and lengthy response times, which raises the overall cost of the procedure (Diasakou, Louloudi, & Papayannakos, 1998). Due to its high price, it is typically not regarded as industrial biodiesel despite the excellent purity of the ester and glycerol (soap-free). Alkaline, acidic, and enzyme catalysts are employed as the three types of catalysts (Thangaraj, Solomon, Muniyandi, Ranganathan, & Lin, 2019). Esters and glycerol are produced as a result of the interaction between a fatty acid or oil and alcohol. There may or may not be a catalyst present in the reaction. High temperature and pressure are necessary for a noncatalytic reaction, and the non-catalytic method has a low yield. The reaction rate and yield are often improved by the catalyst. FFA and alcohol combine in the process to create FAME and water (Murugesan et al., 2009). Excess alcohol alters equilibrium in a forward direction since it is reversible. Transesterification uses primary and secondary alcohols with carbon atoms C_1–C_8 (Formo, 1954). In this situation, the choice of catalyst is likewise quite varied and largely depends on the conversion, corrosion, and separation of the generated end products.

When utilising methanol, typical circumstances are alcohol to vegetable oil molar ratio of 6:1, 60°C–65°C, 1 hour, ambient pressure, and the presence of hydroxide or, more effectively, an alkoxide catalyst (Hoydonckx, De Vos, Chavan, & Jacobs, 2004). The inability of water to form in the reaction $ROH + XOH \rightarrow H_2O + ROX$ (R = alkyl, X = Na or K) has been discovered to make an alkoxide catalyst more effective than hydroxide. Water facilitates the synthesis of FFAs and can lead to incomplete reactions, which can leave monoacylglycerols in the mixture, for example. This means that the feedstock should likewise be devoid of water and FFAs to the greatest extent possible (Hoydonckx et al., 2004). As a result, refined vegetable oil is preferable to crude vegetable oil for the generation of biodiesel. In addition, there are many other transesterification reactions, some of which use enzymatic and heterogeneous catalysts (Rajkumari & Rokhum, 2020; Sahu et al., 2018; Varyambath, Kim, & Kim, 2018). Enzymatic catalysis has the advantages of requiring less or no pre-treatment of feedstocks with high levels of FFAs or low quality, flexibility in the alcohol employed, and high-quality glycerol, but it is more costly and has slower reaction rates. Overall, because this procedure is very mild, quick, and high-yielding, the "traditional" conditions are still essentially in place. Before performing the base-catalysed transesterification reaction, an acid pre-treatment may be necessary if a feedstock with a high acid value, such as discarded frying oil, is employed (Canakci & Van Gerpen, 2001).

6.2.1.3 Advanced biodiesel production methods

In the traditional process of making biodiesel, reactants are heated to a certain temperature to initiate chemical reactions. The most contemporary methods, however, reduce or obtain the necessary activation energy by the use of electricity and other sources. The most current advancements in alternative technologies used in the biodiesel production process were covered in this section.

6.2.1.3.1 Ultrasonic radiation

The esterification of FFAs to FAAEs and the transesterification of TGs to FAAEs are the two main chemical processes used in the synthesis of biodiesel (Fukuda, Kondo, & Noda, 2001). By homogenising the reactants and catalysts to create more microchannels for emulsifying with the reactants and speeding up the reaction rate, ultrasonic chemical radiation aims to boost catalyst surface activation. This reaction mechanism depends on the flow of sonochemical radiations produced by sound waves as well as the acoustic energy they offer. Through the use of piezoelectric transducers, electric energy in ultrasonic radiation is converted to mechanical energy, which is subsequently converted into chemical change (Asakura, Nishida, Matsuoka, & Koda, 2008; Luo et al., 2014). The transesterification and esterification reactions can be greatly enhanced for creating high-quality biodiesel at a low cost and in a significantly shorter amount of time by providing ultrasonic energy with other reaction processes.

6.2.1.3.2 Microwave

The enormous quantity of input energy needed to start the chemical reaction is the fundamental issue with the thermodynamic reaction process. Conversely, because microwave devices can transfer energy directly at the molecular level, they can initiate chemical reactions earlier and use less time and energy while doing so (Soltani et al., 2015). A microwave uses electromagnetic waves that are created through the use of electric energy to transfer energy directly at the molecular level, as opposed to conventional thermal heating, which transfers heat at a high temperature on the surface of reactants before slowly diffusing into the molecule to activate the reaction (Gude, Patil, & Deng, 2012). To achieve complete transesterification reactions, it is crucial to managing the radiation level, as high input can breakdown the compound into several by-products. Facilities for the continuous-flow synthesis of biodiesel using microwave radiation. Several obstacles still exist in the development of industrial-scale microwave-assisted biodiesel reactors, including low product quality and high production costs. According to the research data that is currently available, these issues may be solved by combining microwave radiation with hybrid heating techniques that combine conventional heating with ultrasonic radiation.

6.2.1.3.3 Plasma discharge

As defined by Irving Langmuir in 1932, plasma is an ionised gas made up of electrons, positive and negative ions, and neutral species. Based on temperature and ions, the plasma produced can be divided into two types: cold and hot plasma. When the temperature between the electrons and the ions reaches equilibrium, the hot plasma, or thermal plasma, is produced. This is often accomplished by applying heat and microwave energy at a temperature above 3500°C, creating the ionic phase when the temperature of the electrons and ions are the same (Mostaghimi & Boulos, 2015). Thermal plasma, on the other hand, cannot be utilised to break down the majority of substances because high temperatures can quickly break down even the highly stable bonds in reactants, causing the formation of unnecessary products. Electrons, ions, and radicals are majorly responsible for the non-thermal plasma-induced reaction. The electric field produced by regulating voltages, gas pressure, and power input is largely responsible for the entire response. Catalyst aids in enhancing reaction selectivity and cutting down on reaction time, which is crucial in obtaining the high performance of plasma reactors (Aerts, Somers, & Bogaerts, 2015; Meeprasertsagool, Watthanaphanit, Ueno, Saito, & Reubroycharoen, 2017) for both catalytic and non-catalytic plasma reactions for the synthesis of biodiesel. The biodiesel yield appears to be significantly higher with the catalyst-assisted plasma process than without it. With additional study and development of hybrid plasma technology, it will be possible to use plasma discharge to create any type of chemical reaction in a much more controlled setting. Plasma has a significant impact on catalyst performance, resulting in a substantially longer catalytic life with less coke deposition and catalyst deactivation.

6.2.1.3.4 Supercritical transesterification

A proposed alternative method to current biodiesel production procedures is supercritical transesterification (SCT). Direct TG transesterification and FFA esterification can be accomplished using SCT's catalyst-free method and quick reaction durations, producing a higher-purity end product. The higher phase solubility and minimal mass-transfer restrictions of this catalyst-free process are further benefits (Deshpande, Sunol, & Philippidis, 2017; Farobie & Matsumura, 2017). Although there are numerous advantages to SCT in the production of biodiesel, such as the lack of pre-treatment, the lack of soap, and the removal of catalysts, its main drawback is the high operating expenses caused by the intense temperatures and pressures necessary (Andreo-Martínez et al., 2020). However, a number of modelling, environmental evaluation, and economic studies revealed that supercritical technologies offered superior capital investment, technological advantages, and greater biodiesel generation efficiency with quicker response times with respect to other traditional technologies (Gutiérrez Ortiz, 2020). This approach might be economical at the commercial level, given the higher biodiesel production rate and other raw material costs. Furthermore, energy-saving measures such as utilising the self-heat recuperation technology can help to further increase operation efficiency. The benefits and drawbacks of biodiesel manufacturing technologies as well as the fundamental reaction parameters in each production process, are listed in Table 6.1.

6.2.2 Renewable diesel production methods

The thermochemical and biochemical procedures allow for the generation of RD. While lignocellulosic biomass undergoes the FT process, pyrolysis, and biological conversion of sugars, vegetable oil and animal fats are hydroprocessed or pyrolyzed (Okeke, Sahoo, Kaliyan, & Mani, 2020; SànchezNogué et al., 2018). One method involves hydroprocessing, which involves a number of intricate processes and is done directly on natural TGs to produce RD. Depending on the type of catalyst used and the conditions of the reaction, the three main processes that yield the hydrocarbons $C_{15}-C_{18}$ are HDO, decarboxylation, and decarbonylation (Veriansyah et al., 2012). According to reports, sonochemically treated catalysts convert TGs more effectively than conventionally manufactured catalysts, and high diesel-range hydrocarbon selectivity was seen (Ameen, Azizan, Yusup, & Ramli, 2017). Compared to the transesterification of FAME,

Table 6.1 Advantages and disadvantages of various advanced biodiesel processing technologies.

Technologies	Advantage	Disadvantage
Ultrasonic	• Reduce reaction time and temperature • Low production cost • Improve emulsification of reactants • Reduce reaction temperature, time, alcohol to oil ratio, catalyst amount • Reduce processing time, separation time • Increase yield of biodiesel production by up to 99%	• Difficult to scale up • High energy consumption • Required high-capacity ultrasonic equipment
Microwave	• Efficient heating process • High reaction reactivity • Reduce reaction time • Improve biodiesel yield • Easier separation process and less heat loss • Cleaner products and requires short reaction time	• Quality product • Difficult to scale up • Low heating efficiency • Not efficient for materials with high solids
Plasma	• Reaction at low activation energy • Very fast reaction • Low energy requirement with less catalyst	• Difficult reaction control • Expensive • Specified design reactors
Supercritical	• Fast reaction • No catalyst and environment friendly • Short residence time • No pre-treatment required • Feedstock insensitivity	• High energy consumption • High operating cost • High temperature and pressure disturb biodiesel thermal stability • High alcohol to oil molar ratio

hydroprocessing enables the production of fuel with reduced NOx emissions, improved oxidation stability, and higher energy density (Šimáček, Kubička, Šebor, & Pospíšil, 2010). Additionally, the pyrolysis of feedstocks yields charcoal, non-condensable gases, and bio-oil, the latter of which needs to be upgraded to create RD (Bharath et al., 2021). The limits of bio-oil as a hydrocarbon fuel can be overcome by HDO, catalytic cracking, and supercritical fluids. However, each method has its own limitations, such as the catalytic cracking process's limited catalyst life, HDO's need for high hydrogen pressure, and the price of the solvent used in supercritical fluids (Panwar & Paul, 2021). On the other hand, the FT method is used to produce RD from the syngas produced during the gasification of lignocellulosic biomass. The optimum yield of RD generation and overall energy consumption require the adjustment of the gasifying agent to biomass ratio (Im-orb, Simasatitkul, & Arpornwichanop, 2015). Microorganisms like oleaginous yeast are necessary for the conversion of sugars to RD because they make lipids while consuming the supplied lignocellulosic carbohydrates. The work of SànchezNogué et al. (2018) presented a biological-catalytical hybrid process to yield 31 yeast strains using corn stover hydrolysate. When *Rhodosporidium toruloides* is used as a suitable strain, the generated lipids are extracted, hydrogenated, and isomerised before being converted into RD blend stocks (SànchezNogué et al., 2018). In most cases, catalysts are involved and support the achievement of RD production.

Different second-generation biofuel production methods with substantial potential for large-scale applications that produce various RD fuels develop as the residual biomass feedstock fluctuates. FT diesel is a synthetic diesel created by turning solid biomass into liquids, or "biomass-to-liquids" (Swain, Das, & Naik, 2011). Catalytic hydrotreating of vegetable oils yields paraffinic diesel fuel, commonly known as HVOs, which is how green diesel is made (Kubičková & Kubička, 2010). White diesel is produced via catalytic hydrotreating of plain waste cooking oil or WCO (Bezergianni, Dimitriadis, Kalogianni, & Knudsen, 2011). Hybrid diesel is produced by co-hydro processing of vegetable oils with heavy refinery streams (Bezergianni & Dimitriadis, 2013; Demirbas, 2009).

Catalysts made of one metal or two metals can increase yield and reaction rate. In most cases, monometallic catalysts serve as the parent metal for bimetallic catalysts because they only contain a single type of metal. Ni, Pd, Pt, Rh, Co, Fe, and Ru are the frequently used kinds. It takes a certain type of catalyst to increase the reaction's activity or reduce its output. In order to prevent Ni from causing methanation in the FT process, undesirable side reactions that affect the main reaction pathway due to an insufficient catalyst must be avoided (Martinelli, Gnanamani, LeViness, Jacobs, & Shafer, 2020). Bimetallic catalysts are therefore introduced to address this problem and boost the effectiveness of catalytic processes. Bimetallic catalysts have been shown to be effective in RD production due to their flexibility, toughness, and special features, which monometallic catalysts do not have. Different metal combinations with various supports favour particular reactions. To further improve the catalytic characteristics and activities in boosting catalyst efficiency, more research on bimetallic catalysts for RD production is needed. While conventional bimetallic catalysts have their benefits, additional research on innovative bimetallic catalysts is needed to advance this subject. To fully regulate the final configuration of bimetallic catalysts through synthesis methods and minimise the likelihood of site deactivation, particularly for precious bimetallic catalysts, more research is needed. By improving the stability and recyclability of catalysts, costs for catalysts can be reduced and sustainable, cost-effective, and efficient processes can be achieved in real-world applications.

Numerous articles have reported the laboratory-scale synthesis of renewable fuel. While some papers (Gosling, 2005) discuss the technical-scale production of RD, no information on the nature of the process or the catalysts is provided, probably due to proprietary concerns, with the exception of comments like the mention of using "conventional hydrotreating catalysts" (Stumborg, Wong, & Hogan, 1996). Thus it is challenging to support the claim that this method is less expensive than transesterification (Stumborg et al., 1996). In any event, it seems likely that catalyst systems and environmental factors similar to those found in a lab setting and covered in the following sections are used. According to an implicit statement (Huber, O'Connor, & Corma, 2007), one benefit of RD over biodiesel is that it does not need a separate manufacturing plant because the feedstock can be treated as part of the petrochemical production stream. On the other hand, according to another analysis (Gosling, 2005), building a specific machine for the processing of vegetable oils is more cost-effective. To obtain ultra-low sulphur petrodiesel, HDO is evidently in competition with hydrodesulfurization (Gosling, 2005).

6.3 Properties

6.3.1 Biodiesel properties

Biodiesel is referred to as the alkyl esters of vegetable or animal fats. The fatty acid profile of biodiesel matches that of the parent oil or fat from which it is made. Straight-chain fatty acids, the most prevalent of which have between 16 and 18 carbon atoms, make up the bulk of biodiesel fuels. But in addition to the usual C_{16} and C_{18} acids, some feedstocks may contain sizeable amounts of additional fatty acids. These feedstocks include lauric acid-enriched tropical oils (like coconut oil), which are rich in shorter-chain acids. Minor components of biodiesel include intermediary mono- and diacylglycerols, residual triacylglycerols from the transesterification reaction, residual glycerol and methanol, FFAs, sterols as carryovers from the vegetable oil feedstock, and sterol glucosides. These minor components are in addition to the major fatty ester components. As a result, methanol, acylglycerols, FFAs, and glycerol are restricted in biodiesel standards like ASTM (American Society for Testing and Materials) standard D6751, European standard EN 14214, and other standards being developed globally. IS 15607:2005, a technical standard for biodiesel as a blending stock, was also adopted in India by the Bureau of Indian Standards (BIS). Similar to the current diesel specification, known as IS 1460:2005, is the B5 biodiesel blend specification. BIS is, however, still considering and developing specifications for B6-B20 blends and biodiesel for standalone applications (Bharath Petroleum, 2015). Table 6.2 lists the characteristics of the biodiesel blendstock (IS 15607).

A fuel's qualities are significantly influenced by its composition. The length of the chain and level of unsaturation affect the cetane number of fatty esters. Problems with biodiesel's oxidative stability are caused by the existence of polyunsaturated fatty esters, and cold flow issues are caused by the presence of larger concentrations of saturated fatty esters. Methyl stearate, which has a melting point of about 38°C and oxidative stability of >24 hours as determined by the test EN 14112 (NSAI, 2003) required by the biodiesel standards ASTM D6751 and EN 14214, and methyl linoleate, which has a melting point of 43°C and an oxidative stability of 0.94 hours as determined by the test in EN 14112, serve as examples of this (Table 6.1). As a result, there is a trade-off when trying to address one of these issues by enhancing the fatty ester composition of one type of fatty ester.

Table 6.2 IS 15607 specification of biodiesel (B100) as blending stock (Bharath Petroleum, 2015).

Characteristics	IS 15607:2005
Density, kg/m^3	860–900
Ester content, % min	96.5
Flash point, °C	120
Water and sediment, mg/kg, max.	500
Kinematic viscosity at 40°C, mm^2/s	2.5–6.0
Oxidation stability at 110°C	6

The cloud point (CP) is a crucial factor for biodiesel applications that require low temperatures (Dunn, Shockley, & Bagby, 1996). It is at this temperature that the first visible crystals appear after cooling a fuel, which could lead to issues such as fuel filter clogging. The CP for biodiesel made from soybean oil is close to zero degrees Celsius, and it is only marginally lower for biodiesel made from rapeseed (canola) oil. Due to the presence of saturated fatty esters, biodiesel fuels made from palm oil or animal sources have a substantially higher CP, averaging approximately 15°C. It should be noted that the pour point (PP) is frequently given in relation to low-temperature attributes. The point at which fuel can no longer be poured freely is known as the PP. The cold filter plugging point (CFPP), which is a filterability test and linearly related to the CP, is another test pertaining to cold flow qualities (Dunn et al., 1996).

Research on this topic is still in its infancy (Knothe, 2009; Pinzi et al., 2009). It focuses on improving the qualities of biodiesel fuel by changing and optimising fatty ester composition. Because methyl palmitoleate has a low melting point (about 34°C) (Knothe & Dunn, 2009), palmitoleic acid esters may be acceptable for enrichment in biodiesel. Property enhancement may also result from the enrichment of saturated esters, particularly decanoic acid esters. For instance, the CP of methyl esters of cuphea oil with a fatty acid profile that contains roughly 65% decanoic acid is between 9°C and 10°C (Knothe, Cermak, & Evangelista, 2009). The presence of the ester moiety causes biodiesel to exhibit inherent lubricity (Knothe & Steidley, 2005). Although they are tiny components (impurities) in biodiesel, FFAs and monoacylglycerols significantly increase the lubricity of low-level blends of biodiesel and low-lubricity petrodiesel (1%–2%) biodiesel in petrodiesel. The average flash point of

Table 6.3 Test fuel properties (Na et al., 2015).

Properties	Blend stock	
	An B100 (ASTM D6751)	R100 (D975)
Kinematic viscosity (40°C), mm^2/s	4.4	2.5
Physical distillation, T90, °C, max	348	286
Specific gravity	0.881	0.773
Flash point, °C, min	164	146
Cetane number	57.9	72.3
Cloud point, °C	12.5	-27.1
Sulphur content, ppm, max	2.0	0.3
Nitrogen, ppm	NR	1.3
Carbon (wt.%)	75.9	84.8
Hydrogen (wt.%)	12.2	15.1
Oxygen (wt.%)[a]	11.9	3.0
Total aromatics (wt.%)	NR	0.4

An B, animal-based biodiesel; *NR*, not reported; *R*, renewable diesel.
[a]The highest probable oxygen content is determined by subtracting the sum of weight percentages of Carbon, Hydrogen, and Nitrogen from 100% and assuming that any remaining elements are present in negligible amounts.

biodiesel is higher than that of petroleum diesel. Compared to biodiesel standards, the petrodiesel standards have a lower flash point criterion (Table 6.3). The boiling points of the various constituents are primarily reflected in this. Hexadecane has a boiling temperature of 287°C and decane 174°C, while methyl palmitate has a boiling point of 417°C and methyl decanoate of 224°C (American Chemical Society, 2008). The lower boiling temperatures of the branching and lower molecular weight components of petrodiesel result in a reduction of the flash point. The flash point is affected by the alcohol used in the manufacturing of biodiesel and found in trace amounts in the final product.

6.3.2 Renewable diesel properties

Straight-chain alkanes are the "ideal" constituents of petrodiesel, which is a fairly complex mixture. The high-quality reference compound on the cetane scale, a notion similar to the octane rating used for gasoline and related to the ignition quality of diesel fuel, is hexadecane, sometimes known by the colloquial name "cetane." In addition to these alkanes, branched alkanes and aromatic compounds (with one or more aromatic rings) with alkyl side chains are typically found in petrodiesel. Despite being the low-quality reference chemical on the cetane scale, the highly branched 2,2,4,4,6,8,8-heptamethylnonane has lower melting points, which is an example of how branched compounds, nevertheless, exhibit lower cetane numbers. Alkyl cycloalkanes and alkyl aromatics likewise have low melting temperatures and lower cetane numbers, which rise with side chain length (Clothier, Aguda, Moise, & Pritchard, 1993). In the now-common ULSD fuels, sulphur-containing dibenzothiophenes with alkyl (often methyl) substituents are removed from petrodiesel. Minor oxygen- and nitrogen-containing molecules that had been responsible for petrodiesel's lubricity are also eliminated during the hydrodesulfurization process that results in ULSD fuels (Barbour, Rickeard, & Elliott, 2000). As a result, ULSD fuels have poor lubricity. The lubricity is restored by adding biodiesel at levels of 2% or above, even though lubricity additives (which in some cases may also be based on lipid feedstocks) can also boost lubricity. The purpose of developing RD fuels that mimic petrodiesel is specifically to enrich them with long-chain alkanes, which are, as was already noted, "ideal" components of petrodiesel. RD will have a high cetane number (Table 6.3) if it mostly consists of long-chain alkanes, but if it contains significant amounts of shorter-chain and isomerised species, the cetane number will be lower. It has been reported that RD has a high cetane number (>70) (Aatola, Larmi, Sarjovaara, & Mikkonen, 2009; Kuronen, Mikkonen, Aakko, & Murtonen, 2007; Murtonen et al., 2010). Shorter-chain and isomerised compounds are preferred over long-chain alkanes for better low-temperature characteristics. Although it has typically been reported that the CP of a RD investigated for exhaust emissions testing is between 5°C and 25°C or 30°C, with the precise value depending on the "severity" of the production process or "additional catalytic processing," the CP of this particular diesel was given as 7°C (Aatola et al., 2009; Kuronen et al., 2007) Given that the smaller (and/or branched) molecules required for superior cold flow have lower cetane number, it seems plausible that RD with better cold flow characteristics will exhibit lower cetane number. There appears to be no report on the testing of RD in accordance with the petrodiesel standards ASTM D975 or EN 590 or the jet fuel specifications ASTM D1655 as of yet.

Three blend stocks used to formulate the test fuels, those were: (1) animal-based biodiesel and (2) RD. The gravimetric approach was used to mix-test fuel. In order to prevent oxidation, the headspace of each test fuel drum was filled with inert nitrogen. According to ASTM International Standard ASTM D6751 for Soy B100 and An B100 and D975 for R100 and ULSD, the An B100 and R100 fuels were examined for their qualities. Additionally, the IQT test—which is more precise for high-cetane fuels—was used to assess R100 cetane. Table 6.3 lists the fuels that were examined in detail.

6.4 Combustion characteristics

The combustion characteristics include cylinder pressure, heat release rate, ignition delay (ID), combustion duration, maximum pressure increase rate, brake-specific fuel consumption, and brake thermal efficiency. The combustion characteristics, performance and emissions of a direct injection diesel engine under various operating situations were examined through an experimental enquiry to determine the impacts of adding methanol to biodiesel–diesel mixes. The standard fuel was BD50, which is composed of 50% diesel and 50% biodiesel by volume. Methanol was added to BD50 in amounts of 5% and 10% based on volume (denoted as BDM5 and BDM10). The results show that at low engine load, combustion for BDM5 and BDM10 starts later than for BD50, but at high engine load, combustion is almost equal. The longer ID, better volatility, and lower viscosity of methanol cause a higher volume of fuel to accumulate in the combustion chamber during the premixed burning phase, which results in a high heat release rate even if combustion begins later for the BDM5 and BDM10. BDM5 and BDM10 exhibit similar peak cylinder pressure and peak pressure rise rates to BD50 at low engine load (Pme = 0.177 MPa) and at an engine speed of 1500 rpm and a higher peak heat release rate than BD50. When the engine is running at 1800 rpm, the peak cylinder pressure, peak pressure rise rate, and peak heat release rate of BDM5 and BDM10 are all clearly lower than those of BD50. For BDM5 and BDM10, the peak values occur at crank angles that are later than for BD50. At high engine load (Pme = 0.531 MPa), BDM5 and BDM10 have greater peak cylinder pressures, pressure rise rates, and heat release rates than BD50 and their peak crank angles for heat release rates are nearly identical. The better volatility of methanol, which encourages the formation of a more air–fuel mixture during the premixed burning phase, may be the cause of the increased heat release during premixed burning for BDM5 and BDM10. This indicates that at high engine load, the effect of methanol's reduced cetane number on the ID is diminished.

For all of the tested fuels, the peak cylinder pressure occurs at about the same crank angle under heavy engine loads. The outputs of power and torque from the BDM5 and BDM10 are marginally less than those from the BD50. Smoke emissions have dramatically decreased in BDM5 and BDM10.

6.5 Policy support

International treaties, legislation applicable to commercial energy activities (trading, transport, storage, etc.), financial inducements for investors, standards for usage, conversion and energy production (emission and efficiency standards), taxes and other government tools, promoting energy diversity, general international commerce agreements, energy-related research and innovation, and the green economy are all examples of what constitutes an energy policy in general (Lim & Teong, 2010). Energy policies currently in place also address environmental concerns, such as air pollution, the greenhouse effect (primarily by reducing CO_2 emissions), global warming, and climate change. Eco-friendly technology to boost energy sources and promote cleaner, more efficient energy usage is also included (Hammond, Kallu, & McManus, 2008; Hoekman, 2009; Monni & Raes, 2008; Sawyer, 2009). The growth of the biodiesel business will be directly impacted by energy policy. In particular, through using the right incentives like tax exclusion, price restrictions, targets, and direct subsidies, government plays a vital role in setting the trajectory and, critically, the scale of biodiesel development. The long-term social, economic, and environmental sustainability of bioenergy programmes is a crucial issue. Therefore thorough analyses will be necessary to strengthen their capacity to pinpoint the ideal energy policies that are likely to prevent potential conflicts with food production and land change and offer the greatest benefits (Kaditi, 2009; Phalan, 2009; Pinto et al., 2005). A government can now provide a variety of incentives to encourage the growth of the biodiesel industry and ensure its sustainability, including the planting of crops on fallowed and abandoned agricultural lands, the implementation of a carbon tax, financial support for the production of non-food crops, and an exemption from the oil tax (Ewing & Msangi, 2009; Wassell & Dittmer, 2006). Additionally, several governments have recently made the usage of biodiesel mandatory. For instance, the introduction of the EU's Directive 2003/30/EC mandating the biofuels usage promotion for transportation calls for an increase in the proportion of biofuels from 2% of the overall fuel supply in 2005 to 5.75% of the whole supply in 2010 (based on energy content). Similar regulations have been imposed in other significant nations, such as the United States and Canada, which has resulted in huge demand for biodiesel (Janaun & Ellis, 2010; Mabee, 2007). Additionally, as of now, the general public still has a poor level of understanding of the biodiesel business. The majority of the populace knows very little or nothing about the problems with biodiesel. Governments should therefore improve the promotion of the biodiesel business and raise public understanding of the biodiesel industry, which will aid in the public's support for the government's biodiesel policy (Lim & Teong, 2010). Finally, governments must pay adequate attention to unresolved problems like rainforest destruction and rising food prices while concentrating on strategies to enhance biodiesel production and usage. Around the world, deforestation is thought to be responsible for 20% of greenhouse gas emissions. Additionally, a large portion of the forest currently being cut for palm oil is peatland, which has marshy soils that are essential reservoirs for methane, an even more potent greenhouse gas than carbon dioxide. Their decisions could have a negative impact on climate change if they do not take this into consideration. Governments should encourage biofuel producers to use waste materials, plantations that have received sustainability certification, and crops grown on degraded lands as feedstocks. They should also set clear limits on the amount of fuel that can come from sensitive areas like forests and partially drained peatlands. Governments should provide home farmers with agricultural subsidies so they can maintain a fixed income from growing traditional food crops. In addition, wealthy nations ought to compensate less developed ones to protect forests. By using these methods, plantations can maintain harmony with the forest.

The National Policy on Biofuels, 2018, was amended with approval from the Union Cabinet. The government will now permit more feedstocks for the manufacturing of biofuels, which is a significant amendment. The Ministry of Petroleum and Natural Gas announced the "National Policy on Biofuels in 2018." The National Policy on Biofuels, issued through the Ministry of New & Renewable Energy in 2009, was superseded by the policy, which was announced at the time. The Policy's goal for the next ten years is to increase the use of biofuels in the country's energy and transportation sectors. The Policy intends to utilise, develop, and promote domestic feedstock and its use for the manufacture of biofuels in order to replace fossil fuels more and more, contribute to national energy security, mitigate climate change, and sustainably create new job opportunities. In addition, the policy would promote the use of cutting-edge technologies in the production of biofuels. The Policy classifies biofuels as follows:

- "Basic Biofuels" viz. First Generation (1 G) bioethanol & biodiesel and "Advanced Biofuels."
- "Advance Biofuels" viz. Second Generation (2 G) ethanol, Municipal Solid Waste (MSW) to drop-in fuels.
- Third Generation (3 G) biofuels, bio-CNG etc., to enable the extension of appropriate financial and fiscal incentives under each category.

6.5.1 Biofuel policy in India

- In 2001 India began a tiny 5% ethanol blending trial programme. The National Mission on Biodiesel, which aspired to achieve biodiesel blending of 20% in a diesel by 2011–12, launched in 2003 and established the biofuel mission. The National Policy on Biofuel (NPB) was introduced in 2009 as a result, and the updated NPB was introduced in 2018.
- By 2017, the NPB 2009 anticipated an optional 20% blending objective for both ethanol and biodiesel. Facilitating the best growth of biomass feedstock from domestic waste for biofuel generation was the goal of NPB 2009. The NPB 2009 presented an enabling framework of institutional, financial, and technological interventions in addition to outlining the strategy, vision and goals for biofuel development.
- The updated biofuels policy, NPB 2018, went into effect on May 16, 2018, and it suggested an indicative 2030 blending objective of 20% ethanol in gasoline and 5% biodiesel in diesel. The goal of NPB 2018 is to decrease crude oil imports, increase farmer income, create jobs, make the best use of drylands, and promote sustainability. The goal of the strategy is to offer financial and fiscal incentives that are specific to the first-generation (1 G), second-generation (2 G), and third-generation (3 G) fuels biofuels.

6.6 Challenges and opportunities

Challenges with biodiesel in Asian nations include the price of the feedstock, the availability of feedstock, the effects of the fuel, its viscosity, and the various nations' experiences with nitrogen oxide emissions from the fuel.

6.6.1 Key challenges in biofuel production

6.6.1.1 Feedstock cost

The market for biodiesel had been constrained by expensive feedstock, making it less competitive than petrofuel. Using palm, soybean, and rapeseed oils as the source of biodiesel resulted in production costs of 0.85 USD/L (Indrawan et al., 2017), 0.53 USD/L (You et al., 2008), and 0.705 USD/L (Nejad & Zahedi, 2018), respectively, while the price of producing petrodiesel merely varied from 0.42 USD/L to 0.58 USD/L (Kent, 2018). The price of their feedstocks could have an impact on how much biodiesel and petrodiesel cost to produce. The minimum cost of using palm oil in the years (2015–20) was about 430 USD/tonne, while the maximum cost was about 778 USD/tonne. In the last 5 years, the price of crude oil ranged from 73.3 to 623 USD/tonne. In summary, palm oil was much more expensive on average than crude, which led to a high price for biodiesel. Because of this, the high cost of biodiesel's feedstock had become a concern in Asian nations where palm oil was employed as a fuel source. Because of this, biodiesel was unable to totally replace petroleum-based fuel in the nations. Future research into additional low-cost feedstocks may offer an alternative approach. This necessitates thorough research in this area.

6.6.1.2 Feedstocks supply

Oil palm, soybeans, rapeseed, jatropha etc., oil crops require particular environmental conditions to grow. Locations within 10°N and S of the equator, as well as particular factors like sufficient annual sunlight hours, solar radiation of 16–17 GJ/m^2 daily, high relative humidity daytime, an annual rainfall regime of 2000–2500 mm, and deficit of low vapour pressure is necessary for the growth of oil palm (Comte, Colin, Whalen, Grünberger, & Caliman, 2012). Except for Indonesia, Malaysia, and Thailand, there are not many Asian nations that meet those requirements. As a result, using palm oil as a feedstock for biodiesel appears to be impractical in the majority of Asian nations. In addition, some oil crops, including soybeans, have special planting needs. According to a study, soybeans may be cultivated in several Asian nations where the ideal temperature for growth is between 20°C and 35°C. In China and Japan, especially among undernourished households, biofuels derived from cellulose and agricultural goods, such as soybean, could compete with the demand for food (Koizumi, 2013). Due to the use of waste oil, such as UCO, and animal fats as feedstock, the production of biodiesel in those nations was constrained.

6.6.1.3 Impacts of biodiesel

Globally biodiesel, about 78%, was made from crops in 2017 (UFOP, 2019). In order to meet 20% of the EU's need for biofuels, biofuels, including biodiesel, have led to deforestation in both Indonesia and Malaysia. This demonstrated the potential for first- and second-generation biodiesel to have a detrimental environmental impact. Despite the fact that a lot of crops were grown, their capacity to store carbon may not be as great as that of the forest, which could lead to a rise in carbon dioxide levels and a worsening of global warming. According to a study, widespread oil palm plantations in tropical peatlands could cause ecosystem services, such as the release of carbon dioxide that has been stored, to be lost. As the carbon dioxide in the atmosphere is released, global warming will get worse, which might cause additional environmental issues. Therefore both first- and second-generation biodiesels may have positive and detrimental effects on the environment simultaneously.

6.6.1.4 Viscosity of biodiesel

It is acknowledged that biodiesel frequently has a higher viscosity than petroleum diesel., which could have negative effects on the engine (Hoekman, Broch, Robbins, Ceniceros, & Natarajan, 2012). Reduced atomisation quality was one of the drawbacks of biodiesel caused by its high viscosity (Aldhaidhawi, Chiriac, & Badescu, 2017) and could lead to less heat release rate (Atabani et al., 2012), indicating that petrodiesel provided relatively high energy compared to biodiesel. In this instance, compared to petrodiesel, a greater volume of biodiesel is required to create a given amount of energy, resulting in a low cost–performance ratio. The low cost-to-performance ratio may make matters worse for the general public, who already deemed biodiesel to be prohibitively pricey. Additionally, due to partial combustion, high-viscosity biodiesel may worsen engine-bearing issues in cold weather. Thus the high viscosity of biodiesel turns into a technical obstacle that must be solved in the future development of biodiesel (Alptekin & Canakci, 2008).

6.6.1.5 Nitrogen oxides emission of biodiesel

An analysis of the pertinent literature revealed that biodiesel had greater NOx emissions than petroleum diesel in 85% of the cases studied (Abed, Gad, El Morsi, Sayed, & Elyazeed, 2019). Although it was not generally accepted that biodiesel creates more NOx than petroleum diesel, the considerable number of 85% also implied that biodiesel was likely to emit greater NOx. The relationship between fuel viscosity and NOx emission suggests that higher fuel viscosity could result in higher NOx emission. It is thought that since biodiesel has a greater viscosity than petroleum fuel, it can emit more nitrogen oxides (NOx). The factors that can affect NOx emission were also discovered to include fuel density and cetane number. Fuels with lower cetane numbers and higher fuel densities may produce more NOx emissions (Ban-Weiss, Chen, Buchholz, & Dibble, 2007). NOx is a colourless, odourless gas that can cause major respiratory issues (Turner et al., 2020). Therefore if biodiesel entirely replaces petrofuel in the future, the NOx emission of the fuel must be lowered to avoid such a situation. Solutions like reducing biodiesel's density or viscosity could aid in lowering the fuel's NOx emissions.

6.6.2 Benefits of biodiesel

6.6.2.1 Green technology

Due to its high biodegradability and less toxic nature, biodiesel is considered to be green technology, being environmentally friendly. In the event of a leak or spill, petrodiesel might have a severe negative influence on the environment (Lapinskiene, Martinkus, & Rebždaite, 2006; Leite et al., 2011). In order to determine the sternness of the environmental impacts, toxicity and rate of biodegradability have thus become crucial criteria. In comparison to petroleum diesel, biodiesel could have less of an influence on the environment because it has a higher rate of biodegradation and lower toxicity. A study revealed that whereas petrodiesel only had a 24.5% biodegradation rate after 30 days, different types of biodiesels had similar rates that ranged between 80.4% and 91.2% (Demirbaş, 2009). According to test results, utilising various palm biodiesel blends in a diesel engine reduced emissions of CO and unburnt hydrocarbons by 46% and 73%, respectively, with a slight increase in NOx (Abed et al., 2019).

6.6.2.2 Renewability

The fact that biodiesel is synthesised from both animal and vegetable oils makes it a renewable fuel (Agarwal, 2007). Biodiesel's feedstocks include both edible and non-edible plants, used cooking oil, microalgae, and cyanobacteria, all of which can be cultivated or produced by humans (Abdullah et al., 2019). However, crude oil, a non-renewable feedstock, is the sole source from which petrodiesel (Syafiuddin, Chong, Yuniarto, & Hadibarata, 2020). Crude oil is scarce and will soon run out at the current rate of consumption (Meher, Vidya Sagar, & Naik, 2006). Due to its renewability, biodiesel, therefore, appears to be a good replacement for petroleum diesel.

6.6.2.3 Safety

Biodiesel is comparatively safer than petrodiesel as it has a high flash point. Flash point, the temperature at which a vapour–air mixture can ignite in a closed environment of petrodiesel is lower than that of biodiesel. While petrodiesel exhibits a flash point range lies between 55°C and 66°C, biodiesel has an average flash point of 150°C (Chisti, 2007). Due to weather events, including heat waves, in some Asian nations, the temperature can occasionally reach 40 to 50°C, close to the petrodiesel flash point. In 2019, a heat wave caused several cities in India to reach an average temperature of 45°C, while Jacobabad recorded a temperature of 51.1°C. The future temperature is most likely to rise due to concurrent global warming, which could exacerbate the hazard posed by petrodiesel because of its low flash point and potential for quick ignition. Consequently, biodiesel was safer to store or carry due to its higher flash point.

6.6.2.4 Versatility

It is possible to utilise biodiesel in diesel engines without modifying the hardware. Rudolf Diesel created the first diesel engine in 1897, and it was powered by biodiesel made from vegetable and peanut oils. The efficiency and exhaust emissions of a diesel engine running on biodiesel were investigated through an experiment (Gad et al., 2018). The outcome indicated that utilising biodiesel in a diesel engine had no negative effects other than the fact that its NOx output was a little bit greater than that of petroleum diesel. As a result, it was not necessary to develop a new biodiesel engine, which would save resources and time.

6.6.3 Renewable diesel challenges

- Rising RD demand is causing a growth in the cost of feedstocks such as soyabean oil and animal tallow, thus impacting refiners. The increased demand is also expected to impact the usage of land and increase food prices.

- RD does not have the high energy density of ULSD.
- Developing new logistics systems to supply the waste feedstocks to capable refineries.
- The shortage of raw materials required to accelerate fuel production also acts as a constraint for the industry.
- Meeting global demand forecasts would require diverting raw materials from traditional biodiesel production toward RD production.

6.6.4 Benefits of renewable diesel

- RD significantly reduces greenhouse gas emissions, along with an 80% life cycle reduction in gases and a reduction in carbon emission in diesel engine exhaust by as much as 50%–80% compared to conventional diesel.
- By using a minimum of 14% renewable energy in road and rail transport by 2030, RD is expected to be used as an alternative for refineries to meet Europe's renewable energy directive-II (RED-II) goals (2018).
- Investment in RD can help expand other renewable fuel industries, such as gasoline, naphtha and Sustainable Aviation Fuel (SAF) since they use similar infrastructure and technology.
- Since this diesel meets the ASTM international specifications, it can be processed in existing storage tanks and pipelines.
- It can power conventional auto engines, making it a low-pollution and profitable option.
- The use of RD allows countries to join the circular energy transition without making huge expenses to replace existing infrastructure.

References

Aatola, H., Larmi, M., Sarjovaara, T., & Mikkonen, S. (2009). Hydrotreated vegetable Oil (HVO) as a renewable diesel fuel: Trade-off between NOx, particulate emission, and fuel consumption of a heavy duty engine. *SAE International Journal of Engines, 1*, 1251–1262. Available from https://doi.org/10.4271/2008-01-2500.

Abdullah, B., Syed Muhammad, S. A. F., Shokravi, Z., Ismail, S., Kassim, K. A., Mahmood, A. N., et al. (2019). Fourth generation biofuel: A review on risks and mitigation strategies. *Renewable and Sustainable Energy Reviews, 107*, 37–50. Available from https://doi.org/10.1016/j.rser.2019.02.018.

Abed, K. A., Gad, M. S., El Morsi, A. K., Sayed, M. M., & Elyazeed, S. A. (2019). Effect of biodiesel fuels on diesel engine emissions. *Egyptian Journal of Petroleum, 28*, 183–188. Available from https://doi.org/10.1016/j.ejpe.2019.03.001.

Aerts, R., Somers, W., & Bogaerts, A. (2015). Carbon dioxide splitting in a dielectric barrier discharge plasma: A combined experimental and computational study. *ChemSusChem, 8*, 702–716. Available from https://doi.org/10.1002/cssc.201402818.

Agarwal, A. K. (2007). Biofuels (alcohols and biodiesel) applications as fuels for internal combustion engines. *Progress in Energy and Combustion Science, 33*, 233–271. Available from https://doi.org/10.1016/j.pecs.2006.08.003.

Agrawal, P. S., Belkhode, P. N., & Rokhum, S. L. (2023). *Sustainability in biofuel production technology* (p. 365) John Wiley Sons Ltd.

Aida, B. H. T., Kaouther, Z., Withek, B., Slim, N., & Aymen, O. (2018). Second generation biofuels production from waste cooking oil via pyrolysis process. *Renewable Energy, 126*, 888–896. Available from https://doi.org/10.1016/j.renene.2018.04.002.

Aldhaidhawi, M., Chiriac, R., & Badescu, V. (2017). Ignition delay, combustion and emission characteristics of Diesel engine fueled with rapeseed biodiesel – A literature review. *Renewable and Sustainable Energy Reviews, 73*, 178–186. Available from https://doi.org/10.1016/j.rser.2017.01.129.

Algoufi, Y. T., Akpan, U. G., Kabir, G., Asif, M., & Hameed, B. H. (2017). Upgrading of glycerol from biodiesel synthesis with dimethyl carbonate on reusable Sr–Al mixed oxide catalysts. *Energy Conversion and Management, 138*, 183–189. Available from https://doi.org/10.1016/j.enconman.2017.01.078.

Alleman, T. L., Christensen, E. D., & Moser, B. R. (2019). Improving biodiesel monoglyceride determination by ASTM method D6584-17. *Fuel, 241*, 65–70. Available from https://doi.org/10.1016/j.fuel.2018.12.019.

Alptekin, E., & Canakci, M. (2008). Determination of the density and the viscosities of biodiesel-diesel fuel blends. *Renewable Energy, 33*, 2623–2630. Available from https://doi.org/10.1016/j.renene.2008.02.020.

Al-Saadi, L. S., Eze, V. C., & Harvey, A. P. (2018). Experimental determination of optimal conditions for reactive coupling of biodiesel production with in situ glycerol carbonate formation in a triglyceride transesterification process. *Frontiers in Chemistry, 6*, 1–11. Available from https://doi.org/10.3389/fchem.2018.00625.

Ameen, M., Azizan, M. T., Yusup, S., & Ramli, A. (2017). Hydroprocessing of rubber seed oil over Ni-Mo/γ-Al_2O_3 for the green diesel production. *Chemical Engineering Transactions, 61*, 1843–1848. Available from https://doi.org/10.3303/CET1761305.

American Chemical Society. (2008). Explosives: Completely Revised 6th ed. By Rudolf Meyer (formerly of WASAG Chemie, Essen, Germany), Josef Köhler (Schardenberg, Austria), and Axel Homburg (Dynamit Nobel GmbH, Troisdorf, Germany). Wiley-VCH Verlag GmbH: Weinheim. 2007.viii + 422 pp. $240. ISBN 978-3-527-31656-4. *Journal of American Chemical Society, 130*, 382–384.

Andreo-Martínez, P., Ortiz-Martínez, V. M., García-Martínez, N., de Los Ríos, A. P., Hernández-Fernández, F. J., & Quesada-Medina, J. (2020). Production of biodiesel under supercritical conditions: State of the art and bibliometric analysis. *Applied Energy, 264*, 114753. Available from https://doi.org/10.1016/j.apenergy.2020.114753.

Ao, S., & Rokhum, S. L. (2022). Recent advances in the valorization of biodiesel by-product glycerol to solketal. *Journal of Chemistry, 2022*. Available from https://doi.org/10.1155/2022/4938672.

Arun, N., Sharma, R. V., & Dalai, A. K. (2015). Green diesel synthesis by hydrodeoxygenation of bio-based feedstocks: Strategies for catalyst design and development. *Renewable and Sustainable Energy Reviews, 48*, 240–255. Available from https://doi.org/10.1016/j.rser.2015.03.074.

Asakura, Y., Nishida, T., Matsuoka, T., & Koda, S. (2008). Effects of ultrasonic frequency and liquid height on sonochemical efficiency of large-scale sonochemical reactors. *Ultrasonics Sonochemistry, 15*, 244–250. Available from https://doi.org/10.1016/j.ultsonch.2007.03.012.

Atabani, A. E., Silitonga, A. S., Badruddin, I. A., Mahlia, T. M. I., Masjuki, H. H., & Mekhilef, S. (2012). A comprehensive review on biodiesel as an alternative energy resource and its characteristics. *Renewable and Sustainable Energy Reviews, 16*, 2070–2093. Available from https://doi.org/10.1016/j.rser.2012.01.003.

Badday, A. S., Abdullah, A. Z., Lee, K. T., & Khayoon, M. S. (2012). Intensification of biodiesel production via ultrasonic-assisted process: A critical review on fundamentals and recent development. *Renewable and Sustainable Energy Reviews, 16*, 4574–4587. Available from https://doi.org/10.1016/j.rser.2012.04.057.

Ban-Weiss, G. A., Chen, J. Y., Buchholz, B. A., & Dibble, R. W. (2007). A numerical investigation into the anomalous slight NOx increase when burning biodiesel: A new (old) theory. *Fuel Processing Technology, 88*, 659–667. Available from https://doi.org/10.1016/j.fuproc.2007.01.007.

Barbour, R. H., Rickeard, D. J., & Elliott, N. G. (2000). Understanding diesel lubricity. *SAE Technical Papers, 109*, 1556–1566. Available from https://doi.org/10.4271/2000-01-1918.

Bart, J. C. J., Palmeri, N., & Cavallaro, S. (2010). Vegetable oil formulations for utilisation as biofuels. *Biodiesel Science and Technology*, 114–129. Available from https://doi.org/10.1533/9781845697761.114.

Bashir, M. A., Wu, S., Zhu, J., Krosuri, A., Khan, M. U., & Ndeddy Aka, R. J. (2022). Recent development of advanced processing technologies for biodiesel production: A critical review. *Fuel Processing Technology, 227*, 107120. Available from https://doi.org/10.1016/j.fuproc.2021.107120.

Baskar, G., & Aiswarya, R. (2016). Trends in catalytic production of biodiesel from various feedstocks. *Renewable and Sustainable Energy Reviews, 57*, 496–504. Available from https://doi.org/10.1016/j.rser.2015.12.101.

Bet-Moushoul, E., Farhadi, K., Mansourpanah, Y., Nikbakht, A. M., Molaei, R., & Forough, M. (2016). Application of CaO-based/Au nanoparticles as heterogeneous nanocatalysts in biodiesel production. *Fuel, 164*, 119–127. Available from https://doi.org/10.1016/j.fuel.2015.09.067.

Bezergianni, S., Dimitriadis, A., Kalogianni, A., & Knudsen, K. G. (2011). Toward hydrotreating of waste cooking oil for biodiesel production. Effect of pressure, H_2/oil ratio, and liquid hourly space velocity. *Industrial & Engineering Chemistry Research, 50*, 3874–3879. Available from https://doi.org/10.1021/ie200251a.

Bezergianni, S., & Dimitriadis, A. (2013). Temperature effect on co-hydroprocessing of heavy gas oil-waste cooking oil mixtures for hybrid diesel production. *Fuel, 103*, 579–584. Available from https://doi.org/10.1016/j.fuel.2012.08.006.

Bharath, G., Rambabu, K., Hai, A., Morajkar, P. P., Salkar, A. V., Hasan, S. W., . . . Banat, F., et al. (2021). Highly selective etherification of fructose and 5-hydroxymethylfurfural over a novel Pd-Ru/MXene catalyst for sustainable liquid fuel production. *International Journal of Energy Research, 45*, 14680–14691. Available from https://doi.org/10.1002/er.6743.

Bharath Petroleum. (2015). *Biodiesel: A Clean alternative fuel from renewable resources*. Oper Saf Bull – North.

Buchori, L., Istadi, I., & Purwanto, P. (2016). Advanced chemical reactor technologies for biodiesel production from vegetable oils: A review. *Bulletin of Chemical Reaction Engineering & Catalysis, 11*, 406–430. Available from https://doi.org/10.9767/bcrec.11.3.490.406-430.

Canakci, M., Ozsezen, A. N., Arcaklioglu, E., & Erdil, A. (2009). Prediction of performance and exhaust emissions of a diesel engine fueled with biodiesel produced from waste frying palm oil. *Expert Systems with Applications, 36,* 9268–9280. Available from https://doi.org/10.1016/j.eswa.2008.12.005.

Canakci, M., & Van Gerpen, J. (2001). Biodiesel production from oils and fats with high free fatty acids. *Transactions of the American Society of Agricultural Engineers, 44,* 1429–1436.

Changmai, B., Laskar, I. B., & Rokhum, L. (2019). Microwave-assisted synthesis of glycerol carbonate by the transesterification of glycerol with dimethyl carbonate using Musa acuminata peel ash catalyst. *Journal of the Taiwan Institute of Chemical Engineers, 102,* 276–282. Available from https://doi.org/10.1016/j.jtice.2019.06.014.

Changmai, B., Vanlalveni, C., Ingle, A. P., Bhagat, R., & Rokhum, L. (2020). Widely used catalysts in biodiesel production: A review. *RSC Advances, 10,* 41625–41679. Available from https://doi.org/10.1039/d0ra07931f.

Chen, J. (2017). Techno-economic analysis of biodiesel production from microalgae: A review. *Trends in Renewable Energy, 3,* 141–152. Available from https://doi.org/10.17737/tre.2017.3.2.0035.

Chhandama, M. V. L., Satyan, K. B., Changmai, B., Vanlalveni, C., & Rokhum, S. L. (2021). Microalgae as a feedstock for the production of biodiesel: A review. *Bioresource Technology Reports, 15,* 100771. Available from https://doi.org/10.1016/j.biteb.2021.100771.

Chisti, Y. (2007). Biodiesel from microalgae. *Biotechnology Advances, 25,* 294–306. Available from https://doi.org/10.1016/j.biotechadv.2007.02.001.

Chuah, L. F., Klemeš, J. J., Yusup, S., Bokhari, A., & Akbar, M. M. (2017). A review of cleaner intensification technologies in biodiesel production. *Journal of Cleaner Production, 146,* 181–193. Available from https://doi.org/10.1016/j.jclepro.2016.05.017.

Chuah, L. F., Yusup, S., Abd Aziz, A. R., Bokhari, A., & Abdullah, M. Z. (2016). Cleaner production of methyl ester using waste cooking oil derived from palm olein using a hydrodynamic cavitation reactor. *Journal of Cleaner Production, 112,* 4505–4514. Available from https://doi.org/10.1016/j.jclepro.2015.06.112.

Clothier, P. Q. E., Aguda, B. D., Moise, A., & Pritchard, H. O. (1993). How do diesel-fuel ignition improvers work? *Chemical Society Reviews, 22,* 101–108. Available from https://doi.org/10.1039/CS9932200101.

Comte, I., Colin, F., Whalen, J. K., Grünberger, O., & Caliman, J. P. (2012). *Agricultural practices in oil palm plantations and their impact on hydrological changes,* (1st ed.). *Nutrient fluxes and water quality in Indonesia. A review,* (Vol. 116). Elsevier Inc. Available from https://doi.org/10.1016/B978-0-12-394277-7.00003-8.

Darnoko, D., & Cheryan, M. (2000). Kinetics of palm oil transesterification in a batch reactor. *JAOCS, Journal of the American Oil Chemists' Society, 77,* 1263–1267. Available from https://doi.org/10.1007/s11746-000-0198-y.

Das, A., Shi, D., Halder, G., & Lalthazuala Rokhum, S. (2022). Microwave-assisted synthesis of glycerol carbonate by transesterification of glycerol using Mangifera indica peel calcined ash as catalyst. *Fuel, 330,* 125511. Available from https://doi.org/10.1016/j.fuel.2022.125511.

Das, A., Shi, D., Halder, G., & Lalthazuala, S. (2022). Microwave-assisted synthesis of glycerol carbonate by transesterification of glycerol using Mangifera indica peel calcined ash as catalyst. *Fuel, 330,* 125511. Available from https://doi.org/10.1016/j.fuel.2022.125511.

de Jesus, A., Silva, M. M., & Vale, M. G. R. (2008). The use of microemulsion for determination of sodium and potassium in biodiesel by flame atomic absorption spectrometry. *Talanta, 74*, 1378–1384. Available from https://doi.org/10.1016/j.talanta.2007.09.010.

Demirbaş, A. (2009). Biodegradability of biodiesel and petrodiesel fuels. *Energy Sources, Part A: Recovery, Utilization and Environmental Effects, 31*, 169–174. Available from https://doi.org/10.1080/15567030701521809.

Demirbas, A. (2009). Progress and recent trends in biodiesel fuels. *Energy Conversion and Management, 50*, 14–34. Available from https://doi.org/10.1016/j.enconman.2008.09.001.

Deshpande, S. R., Sunol, A. K., & Philippidis, G. (2017). Status and prospects of supercritical alcohol transesterification for biodiesel production. *Wiley Interdisciplinary Reviews: Energy and Environment, 6*, 1–15. Available from https://doi.org/10.1002/wene.252.

Diasakou, M., Louloudi, A., & Papayannakos, N. (1998). Kinetics of the non-catalytic transesterification of soybean oil. *Fuel, 77*, 1297–1302. Available from https://doi.org/10.1016/S0016-2361(98)00025-8.

Ding, X., Yuan, X., Leng, L., Huang, H., Wang, H., Shao, J., et al. (2017). Upgrading sewage sludge liquefaction bio-oil by microemulsification: The effect of ethanol as polar phase on solubilization performance and fuel properties. *Energy and Fuels, 31*, 1574–1582. Available from https://doi.org/10.1021/acs.energyfuels.6b02269.

Dunn, R. O., Shockley, M. W., & Bagby, M. O. (1996). Improving the low-temperature properties of alternative diesel fuels: Vegetable oil-derived methyl esters. *JAOCS, Journal of the American Oil Chemists' Society, 73*, 1719–1728. Available from https://doi.org/10.1007/BF02517978.

European Committee for Standardization. (2014). *EN 14214 Fatty acid methyl esters (FAME) Requirements and test methods* (p. 10).

European Standard. (2013). EN 590:2013. Automotive fuels – diesel – requirements and test methods (pp. 1–12).

Ewing, M., & Msangi, S. (2009). Biofuels production in developing countries: Assessing tradeoffs in welfare and food security. *Environmental Science & Policy, 12*, 520–528. Available from https://doi.org/10.1016/j.envsci.2008.10.002.

Farobie, O., & Matsumura, Y. (2017). State of the art of biodiesel production under supercritical conditions. *Progress in Energy and Combustion Science, 63*, 173–203. Available from https://doi.org/10.1016/j.pecs.2017.08.001.

Formo, M. W. (1954). Ester reactions of fatty materials. *Journal of the American Oil Chemists' Society, 31*, 548–559. Available from https://doi.org/10.1007/BF02638571.

Fukuda, H., Kondo, A., & Noda, H. (2001). Biodiesel fuel production by transesterification of oils. *Journal of Bioscience and Bioengineering, 92*, 405–416. Available from https://doi.org/10.1016/S1389-1723(01)80288-7.

Gad, M. S., El-Araby, R., Abed, K. A., El-Ibiari, N. N., El Morsi, A. K., & El-Diwani, G. I. (2018). Performance and emissions characteristics of C.I. engine fueled with palm oil/palm oil methyl ester blended with diesel fuel. *Egyptian Journal of Petroleum, 27*, 215–219. Available from https://doi.org/10.1016/j.ejpe.2017.05.009.

Gosling, C. (2005). A technoeconomic and environmental life cycle comparison of green diesel to biodiesel and Syndiesel: A numerical investigation into the anomalous slight NOx increase when burning biodiesel: A new (old) theory. *SAE Technical Papers*. Available from https://doi.org/10.4271/2005-01-3771.

Gude, V. G., Patil, P., & Deng, S. (2012). *Microwave energy potential for large scale biodiesel production*. In World Renew Energy Forum, WREF 2012, Incl World

Renew Energy Congr XII Color Renew Energy Soc Annu Conf, Vol. 1, pp. 751–759.

Gulum, M., Kutlu Onay, F., & Bilgin, A. (2018). Evaluation of predictive capabilities of regression models and artificial neural networks for density and viscosity measurements of different biodiesel-diesel-vegetable oil ternary blends. *Environmental and Climate Technologies, 22*, 179–205. Available from https://doi.org/10.2478/rtuect-2018-0012.

Gutiérrez Ortiz, F. J. (2020). Techno-economic assessment of supercritical processes for biofuel production. *The Journal of Supercritical Fluids, 160*, 1–15. Available from https://doi.org/10.1016/j.supflu.2020.104788.

Hajbabaei, M., Johnson, K. C., Okamoto, R. A., Mitchell, A., Pullman, M., & Durbin, T. D. (2012). Evaluation of the impacts of biodiesel and second generation biofuels on NOx emissions for CARB diesel fuels. *Environmental Science & Technology, 46*, 9163–9173. Available from https://doi.org/10.1021/es300739r.

Hammond, G. P., Kallu, S., & McManus, M. C. (2008). Development of biofuels for the UK automotive market. *Applied Energy, 85*, 506–515. Available from https://doi.org/10.1016/j.apenergy.2007.09.005.

Higman, E. B., Higman, H. C., Chortyk, O. T., & Schmeltz, I. (1973). Studies on the thermal degradation of naturally occurring materials. II. Products from the pyrolysis of triglycerides at 400 degrees. *Journal of Agricultural and Food Chemistry, 21*, 202–204. Available from https://doi.org/10.1021/jf60186a030.

Hoekman, S. K. (2009). Biofuels in the U.S. – challenges and opportunities. *Renewable Energy, 34*, 14–22. Available from https://doi.org/10.1016/j.renene.2008.04.030.

Hoekman, S. K., Broch, A., Robbins, C., Ceniceros, E., & Natarajan, M. (2012). Review of biodiesel composition, properties, and specifications. *Renewable and Sustainable Energy Reviews, 16*, 143–169. Available from https://doi.org/10.1016/j.rser.2011.07.143.

Hoydonckx, H. E., De Vos, D. E., Chavan, S. A., & Jacobs, P. A. (2004). Esterification and transesterification of renewable chemicals. *Topics in Catalysis, 27*, 83–96. Available from https://doi.org/10.1023/B:TOCA.0000013543.96438.1a.

Huber, G. W., & Corma, A. (2007). Synergies between bio- and oil refineries for the production of fuels from biomass. *Angewandte Chemie International Edition, 46*, 7184–7201. Available from https://doi.org/10.1002/anie.200604504.

Huber, G. W., O'Connor, P., & Corma, A. (2007). Processing biomass in conventional oil refineries: Production of high quality diesel by hydrotreating vegetable oils in heavy vacuum oil mixtures. *Applied Catalysis A: General, 329*, 120–129. Available from https://doi.org/10.1016/j.apcata.2007.07.002.

Ibrahim, N. A., Rashid, U., Hazmi, B., Moser, B. R., Alharthi, F. A., Rokhum, S. L., et al. (2022). Biodiesel production from waste cooking oil using magnetic bifunctional calcium and iron oxide nanocatalysts derived from empty fruit bunch. *Fuel, 317*, 123525. Available from https://doi.org/10.1016/j.fuel.2022.123525.

Im-orb, K., Simasatitkul, L., & Arpornwichanop, A. (2015). *Performance Analysis and Optimization of the Biomass Gasification and Fischer–Tropsch Integrated Process for Green Fuel Productions* (Vol. 37). Elsevier. Available from https://doi.org/10.1016/B978-0-444-63578-5.50041-4.

Indrawan, N., Thapa, S., Rahman, S. F., Park, J. H., Park, S. H., Wijaya, M. E., et al. (2017). Palm biodiesel prospect in the Indonesian power sector.

Environmental Technology & Innovation, 7, 110–127. Available from https://doi.org/10.1016/j.eti.2017.01.001.

Janaun, J., & Ellis, N. (2010). Perspectives on biodiesel as a sustainable fuel. *Renewable and Sustainable Energy Reviews, 14*, 1312–1320. Available from https://doi.org/10.1016/j.rser.2009.12.011.

Kaditi, E. A. (2009). Bio-energy policies in a global context. *Journal of Cleaner Production, 17*, S4–S8. Available from https://doi.org/10.1016/j.jclepro.2008.08.023.

Kalsoom, M., Zerey-belaskri, A. El, Nadeem, F., & Shehzad, M. R. (2017). Fatty acid chain length optimization for biodiesel production using different chemical and biochemical approaches – a comprehensive review. *International Journal of Chemical and Biochemical Sciences, 11*, 75–94.

Kent, R. (2018). Renewables. *Plastics Engineering, 74*, 56–57. Available from https://doi.org/10.1002/peng.20026.

Kim, D. S., Hanifzadeh, M., & Kumar, A. (2018). Trend of biodiesel feedstock and its impact on biodiesel emission characteristics. *Environmental Progress & Sustainable Energy, 37*, 7–19. Available from https://doi.org/10.1002/ep.12800.

Knothe, G. (2009). Improving biodiesel fuel properties by modifying fatty ester composition. *Energy & Environmental Science, 2*, 759–766. Available from https://doi.org/10.1039/b903941d.

Knothe, G. (2010). Biodiesel and renewable diesel: A comparison. *Progress in Energy and Combustion Science, 36*, 364–373. Available from https://doi.org/10.1016/j.pecs.2009.11.004.

Knothe, G., Cermak, S. C., & Evangelista, R. L. (2009). Cuphea oil as source of biodiesel with improved fuel properties caused by high content of methyl decanoate. *Energy and Fuels, 23*, 1743–1747. Available from https://doi.org/10.1021/ef800958t.

Knothe, G., & Dunn, R. O. (2009). A comprehensive evaluation of the melting points of fatty acids and esters determined by differential scanning calorimetry. *JAOCS, Journal of the American Oil Chemists' Society, 86*, 843–856. Available from https://doi.org/10.1007/s11746-009-1423-2.

Knothe, G., & Steidley, K. R. (2005). Lubricity of components of biodiesel and petrodiesel. The origin of biodiesel lubricity. *Energy and Fuels, 19*, 1192–1200. Available from https://doi.org/10.1021/ef049684c.

Koizumi, T. (2013). Biofuel and food security in China and Japan. *Renewable and Sustainable Energy Reviews, 21*, 102–109. Available from https://doi.org/10.1016/j.rser.2012.12.047.

Kubičková, I., & Kubička, D. (2010). Utilization of triglycerides and related feedstocks for production of clean hydrocarbon fuels and petrochemicals: A review. *Waste and Biomass Valorization, 1*, 293–308. Available from https://doi.org/10.1007/s12649-010-9032-8.

Kumawat, M. K., & Rokhum, S. L. (2022). Biodiesel production from oleic acid using biomass-derived sulfonated orange peel catalyst. *Frontiers in Catalysis, 2*, 1–12. Available from https://doi.org/10.3389/fctls.2022.914670.

Kuronen, M., Mikkonen, S., Aakko, P., & Murtonen, T. (2007). Hydrotreated vegetable oil as fuel for heavy duty diesel engines. *SAE Technical Papers*. Available from https://doi.org/10.4271/2007-01-4031.

Lapinskiene, A., Martinkus, P., & Rebždaite, V. (2006). Eco-toxicological studies of diesel and biodiesel fuels in aerated soil. *Environmental Pollution (Barking, Essex: 1987), 142*, 432–437. Available from https://doi.org/10.1016/j.envpol.2005.10.023.

Laskar, I. B., Rajkumari, K., Gupta, R., Chatterjee, S., Paul, B., & Rokhum, L. (2018). Waste snail shell derived heterogeneous catalyst for biodiesel production by the transesterification of soybean oil. *RSC Advances, 8,* 20131–20142. Available from https://doi.org/10.1039/c8ra02397b.

Leite, M. B. N. L., de Araújo, M. M. S., Nascimento, I. A., da Cruz, A. C. S., Pereira, S. A., & do Nascimento, N. C. (2011). Toxicity of water-soluble fractions of biodiesel fuels derived from castor oil, palm oil, and waste cooking oil. *Environmental Toxicology and Chemistry/SETAC, 30,* 893–897. Available from https://doi.org/10.1002/etc.444.

Leng, L., Han, P., Yuan, X., Li, J., & Zhou, W. (2018). Biodiesel microemulsion upgrading and thermogravimetric study of bio-oil produced by liquefaction of different sludges. *Energy, 153,* 1061–1072. Available from https://doi.org/10.1016/j.energy.2018.04.087.

Li, M., Liu, S., Li, S., Yang, Y., Cui, Y., & Gong, J. (2018). Determination and correlation of dipyridamole p-toluene sulfonate solubility in seven alcohol solvents and three binary solvents. *Journal of Chemical & Engineering Data, 63,* 208–216. Available from https://doi.org/10.1021/acs.jced.7b00825.

Likozar, B., & Levec, J. (2014). Transesterification of canola, palm, peanut, soybean and sunflower oil with methanol, ethanol, isopropanol, butanol and tert-butanol to biodiesel: Modelling of chemical equilibrium, reaction kinetics and mass transfer based on fatty acid composition. *Applied Energy, 123,* 108–120. Available from https://doi.org/10.1016/j.apenergy.2014.02.046.

Lim, S., & Teong, L. K. (2010). Recent trends, opportunities and challenges of biodiesel in Malaysia: An overview. *Renewable and Sustainable Energy Reviews, 14,* 938–954. Available from https://doi.org/10.1016/j.rser.2009.10.027.

Luo, J., Fang, Z., & Smith, R. L. (2014). Ultrasound-enhanced conversion of biomass to biofuels. *Progress in Energy and Combustion Science, 41,* 56–93. Available from https://doi.org/10.1016/j.pecs.2013.11.001.

Ma, F., & Hanna, M. A. (1999). Biodiesel production: A review1. *Bioresource Technology, 70,* 1–15. Available from https://doi.org/10.1016/s0960-8524(99)00025-5.

Mabee, W. E. (2007). Policy options to support biofuel production. *Advances in Biochemical Engineering/Biotechnology, 108,* 329–357. Available from https://doi.org/10.1007/10_2007_059.

Martinelli, M., Gnanamani, M. K., LeViness, S., Jacobs, G., & Shafer, W. D. (2020). An overview of Fischer–Tropsch Synthesis: XtL processes, catalysts and reactors. *Applied Catalysis A: General, 608,* 117740. Available from https://doi.org/10.1016/j.apcata.2020.117740.

Meeprasertsagool, P., Watthanaphanit, A., Ueno, T., Saito, N., & Reubroycharoen, P. (2017). New insights into vegetable oil pyrolysis by cold plasma technique. *Energy Procedia, 138,* 1153–1158. Available from https://doi.org/10.1016/j.egypro.2017.10.224.

Meher, L. C., Vidya Sagar, D., & Naik, S. N. (2006). Technical aspects of biodiesel production by transesterification: A review. *Renewable and Sustainable Energy Reviews, 10,* 248–268. Available from https://doi.org/10.1016/j.rser.2004.09.002.

Monni, S., & Raes, F. (2008). Multilevel climate policy: The case of the European Union, Finland and Helsinki. *Environmental Science & Policy, 11,* 743–755. Available from https://doi.org/10.1016/j.envsci.2008.08.001.

Mostaghimi, J., & Boulos, M. I. (2015). Thermal plasma sources: How well are they adopted to process needs. *Plasma Chemistry and Plasma Processing, 35,* 421–436. Available from https://doi.org/10.1007/s11090-015-9616-y.

Motasemi, F., & Ani, F. N. (2012). A review on microwave-assisted production of biodiesel. *Renewable and Sustainable Energy Reviews, 16*, 4719–4733. Available from https://doi.org/10.1016/j.rser.2012.03.069.

Mueanmas, C., Nikhom, R., Petchkaew, A., Iewkittayakorn, J., & Prasertsit, K. (2019). Extraction and esterification of waste coffee grounds oil as non-edible feedstock for biodiesel production. *Renewable Energy, 133*, 1414–1425. Available from https://doi.org/10.1016/j.renene.2018.08.102.

Muniyappa, P. R., Brammer, S. C., & Noureddini, H. (1996). Improved conversion of plant oils and animal fats into biodiesel and co-product. *Bioresource Technology, 56*, 19–24. Available from https://doi.org/10.1016/0960-8524(95)00178-6.

Murtonen, A. T., Kuronen, M., Mikkonen, S., Lehtoranta, K., Sae, S., Journal, I., et al. (2010). *Emissions with heavy-duty diesel engines and vehicles using FAME, HVO and GTL fuels with and without DOC + POC Aftertreatment. SAE International Journal of Fuels and Lubricants* (2, pp. 147–166).

Murugesan, A., Umarani, C., Chinnusamy, T. R., Krishnan, M., Subramanian, R., & Neduzchezhain, N. (2009). Production and analysis of bio-diesel from non-edible oils: A review. *Renewable and Sustainable Energy Reviews, 13*, 825–834. Available from https://doi.org/10.1016/j.rser.2008.02.003.

Mustranta, A., Forssell, P., & Poutanen, K. (1993). Applications of immobilized lipases to transesterification and esterification reactions in nonaqueous systems. *Enzyme and Microbial Technology, 15*, 133–139. Available from https://doi.org/10.1016/0141-0229(93)90037-3.

Na, K., Biswas, S., Robertson, W., Sahay, K., Okamoto, R., Mitchell, A., et al. (2015). Impact of biodiesel and renewable diesel on emissions of regulated pollutants and greenhouse gases on a 2000 heavy duty diesel truck. *Atmospheric Environment (Oxford, England: 1994), 107*, 307–314. Available from https://doi.org/10.1016/j.atmosenv.2015.02.054.

Najafi, B., Akbarian, E., Lashkarpour, S. M., Aghbashlo, M., Ghaziaskar, H. S., & Tabatabaei, M. (2019). Modeling of a dual fueled diesel engine operated by a novel fuel containing glycerol triacetate additive and biodiesel using artificial neural network tuned by genetic algorithm to reduce engine emissions. *Energy, 168*, 1128–1137. Available from https://doi.org/10.1016/j.energy.2018.11.142.

Navas, M. B., Lick, I. D., Bolla, P. A., Casella, M. L., & Ruggera, J. F. (2018). Transesterification of soybean and castor oil with methanol and butanol using heterogeneous basic catalysts to obtain biodiesel. *Chemical Engineering Science, 187*, 444–454. Available from https://doi.org/10.1016/j.ces.2018.04.068.

Ndiaye, M., Arhaliass, A., Legrand, J., Roelens, G., & Kerihuel, A. (2020). Reuse of waste animal fat in biodiesel: Biorefining heavily-degraded contaminant-rich waste animal fat and formulation as diesel fuel additive. *Renewable Energy, 145*, 1073–1079. Available from https://doi.org/10.1016/j.renene.2019.06.030.

Nejad, A. S., & Zahedi, A. R. (2018). Optimization of biodiesel production as a clean fuel for thermal power plants using renewable energy source. *Renewable Energy, 119*, 365–374. Available from https://doi.org/10.1016/j.renene.2017.11.087.

Neste Corporation. (2015). *Neste renewable diesel handbook* (pp. 1–33). Neste.

Nigam, P. S., & Singh, A. (2011). Production of liquid biofuels from renewable resources. *Progress in Energy and Combustion Science, 37*, 52–68. Available from https://doi.org/10.1016/j.pecs.2010.01.003.

NSAI. (2003). *Standard fat and oil derivatives – fatty acid methyl esters (FAME) – determination of oxidation stability.* Natl Stand Auth Irel.

Ogunkunle, O., & Ahmed, N. A. (2019). A review of global current scenario of biodiesel adoption and combustion in vehicular diesel engines. *Energy*

Reports, 5, 1560–1579. Available from https://doi.org/10.1016/j.egyr.2019.10.028.

Okeke, I. J., Sahoo, K., Kaliyan, N., & Mani, S. (2020). Life cycle assessment of renewable diesel production via anaerobic digestion and Fischer–Tropsch synthesis from miscanthus grown in strip-mined soils. *Journal of Cleaner Production, 249*, 119358. Available from https://doi.org/10.1016/j.jclepro.2019.119358.

Pan, H., Xia, Q., Wang, Y., Shen, Z., Huang, H., Ge, Z., et al. (2022). Recent advances in biodiesel production using functional carbon materials as acid/base catalysts. *Fuel Processing Technology, 237*, 107421. Available from https://doi.org/10.1016/j.fuproc.2022.107421.

Panwar, N. L., & Paul, A. S. (2021). An overview of recent development in bio-oil upgrading and separation techniques. *Environmental Engineering Research, 26*, 200382. Available from https://doi.org/10.4491/eer.2020.382.

Pathak, G., Das, D., Rajkumari, K., & Rokhum, L. (2018). Exploiting waste: Towards a sustainable production of biodiesel using: Musa acuminata peel ash as a heterogeneous catalyst. *Green Chemistry: An International Journal and Green Chemistry Resource: GC, 20*, 2365–2373. Available from https://doi.org/10.1039/c8gc00071a.

Phalan, B. (2009). The social and environmental impacts of biofuels in Asia: An overview. *Applied Energy, 86*, S21–S29. Available from https://doi.org/10.1016/j.apenergy.2009.04.046.

Pinto, A. C., Guarieiro, L. L. N., Rezende, M. J. C., Ribeiro, N. M., Torres, E. A., Lopes, W. A., et al. (2005). Biodiesel: An overview. *Journal of the Brazilian Chemical Society, 16*, 1313–1330. Available from https://doi.org/10.1590/S0103-50532005000800003.

Pinzi, S., Garcia, I. L., Lopez-Gimenez, F. J., DeCastro, M. D. L., Dorado, G., & Dorado, M. P. (2009). The ideal vegetable oil-based biodiesel composition: A review of social, economical and technical implications. *Energy and Fuels, 23*, 2325–2341. Available from https://doi.org/10.1021/ef801098a.

Pölczmann, G., Tóth, O., Beck, Á., & Hancsók, J. (2016). Investigation of storage stability of diesel fuels containing biodiesel produced from waste cooking oil. *Journal of Cleaner Production, 111*, 85–92. Available from https://doi.org/10.1016/j.jclepro.2015.08.035.

Rajkumari, K., Changmai, B., Meher, A. K., Vanlalveni, C., Sudarsanam, P., Wheatley, A. E. H., et al. (2021). A reusable magnetic nanocatalyst for bio-fuel additives: The ultrasound-assisted synthesis of solketal. *Sustainable Energy & Fuels, 5*, 2362–2372. Available from https://doi.org/10.1039/d0se01900c.

Rajkumari, K., & Rokhum, L. (2020). A sustainable protocol for production of biodiesel by transesterification of soybean oil using banana trunk ash as a heterogeneous catalyst. *Biomass Conversion & Biorefinery, 10*, 839–848. Available from https://doi.org/10.1007/s13399-020-00647-8.

Regulation, A. (2010). *Adopted regulation of the state board of agriculture LCB file no. R209-08* (pp. 1–7). Effective January 28, 2010.

REN21. (2017). Global status report. *Renewable 2020, 72*.

Rezania, S., Oryani, B., Park, J., Hashemi, B., Yadav, K. K., Kwon, E. E., et al. (2019). Review on transesterification of non-edible sources for biodiesel production with a focus on economic aspects, fuel properties and by-product applications. *Energy Conversion and Management, 201*, 112155. Available from https://doi.org/10.1016/j.enconman.2019.112155.

Sahu, G., Gupta, N. K., Kotha, A., Saha, S., Datta, S., Chavan, P., et al. (2018). A review on biodiesel production through heterogeneous catalysis route.

ChemBioEng Reviews, *5*, 231–252. Available from https://doi.org/10.1002/cben.201700014.

Saikia, K., Rajkumari, K., Moyon, N. S., Basumatary, S., Halder, G., Rashid, U., et al. (2022). Sulphonated biomass-based catalyst for solketal synthesis by acetalization of glycerol – a byproduct of biodiesel production. *Fuel Processing Technology*, *238*, 107482. Available from https://doi.org/10.1016/j.fuproc.2022.107482.

SànchezNogué, V., Black, B. A., Kruger, J. S., Singer, C. A., Ramirez, K. J., Reed, M. L., et al. (2018). Integrated diesel production from lignocellulosic sugars: Via oleaginous yeast. *Green Chemistry: An International Journal and Green Chemistry Resource: GC*, *20*, 4349–4365. Available from https://doi.org/10.1039/c8gc01905c.

Sawyer, R. F. (2009). Science based policy for addressing energy and environmental problems. *Proceedings of the Combustion Institute*, *32*(1), 45–56. Available from https://doi.org/10.1016/j.proci.2008.07.003.

Schwab, A. W., Dykstra, G. J., Selke, E., Sorenson, S. C., & Pryde, E. H. (1988). Diesel fuel from thermal decomposition of soybean oil. *Journal of the American Oil Chemists' Society*, *65*, 1781–1786. Available from https://doi.org/10.1007/BF02542382.

Semorile, N. F., Alviso, D., & Romano, S. D. (2023). Flash point and refractive index measurements of diesel and biodiesel, and their binary blends with n-butanol and n-pentanol. *Journal of the Brazilian Society of Mechanical Sciences and Engineering*, *45*. Available from https://doi.org/10.1007/s40430-022-03943-8.

Shuit, S. H., Ong, Y. T., Lee, K. T., Subhash, B., & Tan, S. H. (2012). Membrane technology as a promising alternative in biodiesel production: A review. *Biotechnology Advances*, *30*, 1364–1380. Available from https://doi.org/10.1016/j.biotechadv.2012.02.009.

Silitonga, A. S., Shamsuddin, A. H., Mahlia, T. M. I., Milano, J., Kusumo, F., Siswantoro, J., et al. (2020). Biodiesel synthesis from Ceiba pentandra oil by microwave irradiation-assisted transesterification: ELM modeling and optimization. *Renewable Energy*, *146*, 1278–1291. Available from https://doi.org/10.1016/j.renene.2019.07.065.

Šimáček, P., Kubička, D., Šebor, G., & Pospíšil, M. (2010). Fuel properties of hydroprocessed rapeseed oil. *Fuel*, *89*, 611–615. Available from https://doi.org/10.1016/j.fuel.2009.09.017.

Siva, R., Munuswamy, D. B., & Devarajan, Y. (2019). Emission and performance study emulsified orange peel oil biodiesel in an aspirated research engine. *Petroleum Science*, *16*, 180–186. Available from https://doi.org/10.1007/s12182-018-0288-0.

Soltani, S., Rashid, U., Yunus, R., & Taufiq-Yap, Y. H. (2015). Synthesis of biodiesel through catalytic transesterification of various feedstocks using fast solvothermal technology: A critical review. *Catalysis Reviews – Science and Engineering*, *57*, 407–435. Available from https://doi.org/10.1080/01614940.2015.1066640.

Srifa, A., Faungnawakij, K., Itthibenchapong, V., & Assabumrungrat, S. (2015). Roles of monometallic catalysts in hydrodeoxygenation of palm oil to green diesel. *Chemical Engineering Journal*, *278*, 249–258. Available from https://doi.org/10.1016/j.cej.2014.09.106.

Stumborg, M., Wong, A., & Hogan, E. (1996). Hydroprocessed vegetable oils for diesel fuel improvement. *Bioresource Technology*, *56*, 13–18. Available from https://doi.org/10.1016/0960-8524(95)00181-6.

Sundus, F., Fazal, M. A., & Masjuki, H. H. (2017). Tribology with biodiesel: A study on enhancing biodiesel stability and its fuel properties. *Renewable and*

Sustainable Energy Reviews, 70, 399–412. Available from https://doi.org/10.1016/j.rser.2016.11.217.

Swain, P. K., Das, L. M., & Naik, S. N. (2011). Biomass to liquid: A prospective challenge to research and development in 21st century. *Renewable and Sustainable Energy Reviews, 15*, 4917–4933. Available from https://doi.org/10.1016/j.rser.2011.07.061.

Syafiuddin, A., Chong, J. H., Yuniarto, A., & Hadibarata, T. (2020). The current scenario and challenges of biodiesel production in Asian countries: A review. *Bioresource Technology Reports, 12*, 100608. Available from https://doi.org/10.1016/j.biteb.2020.100608.

Tan, K. T., & Lee, K. T. (2011). A review on supercritical fluids (SCF) technology in sustainable biodiesel production: Potential and challenges. *Renewable and Sustainable Energy Reviews, 15*, 2452–2456. Available from https://doi.org/10.1016/j.rser.2011.02.012.

Tan, Y. L., Abdullah, A. Z., & Hameed, B. H. (2018). Catalytic fast pyrolysis of durian rind using silica-alumina catalyst: Effects of pyrolysis parameters. *Bioresource Technology, 264*, 198–205. Available from https://doi.org/10.1016/j.biortech.2018.05.058.

Thangaraj, B., Solomon, P. R., Muniyandi, B., Ranganathan, S., & Lin, L. (2019). Catalysis in biodiesel production: A review. *Clean Energy, 3*, 2–23. Available from https://doi.org/10.1093/ce/zky020.

Trejo-Zárraga, F., Hernández-Loyo, F. d. J., Chavarría-Hernández, J. C., & Sotelo-Boyás, R. (2018). Kinetics of transesterification processes for biodiesel production. *Biofuels - State of Development*. Available from https://doi.org/10.5772/intechopen.75927.

Troter, D. Z., Todorović, Z. B., Dokić-Stojanović, D. R., Stamenković, O. S., & Veljković, V. B. (2016). Application of ionic liquids and deep eutectic solvents in biodiesel production: A review. *Renewable and Sustainable Energy Reviews, 61*, 473–500. Available from https://doi.org/10.1016/j.rser.2016.04.011.

Turner, M. C., Andersen, Z. J., Baccarelli, A., Diver, W. R., Gapstur, S. M., Pope, C. A., et al. (2020). Outdoor air pollution and cancer: An overview of the current evidence and public health recommendations. *CA: A Cancer Journal for Clinicians, 70*, 460–479. Available from https://doi.org/10.3322/caac.21632.

UFOP. (2019). UFOP report on global market supply: 2017/2018 (p. 51).

Van Gerpen, J., Shanks, B., Pruszko, R., Clements, D., & Knothe, G. (2004). *Biodiesel analytical methods August 2002–January 2004* (p. 100). Nrel/Sr-510-36240.

Varyambath, A., Kim, M. R., & Kim, I. (2018). Sulfonic acid-functionalized organic knitted porous polyaromatic microspheres as heterogeneous catalysts for biodiesel production. *New Journal of Chemistry, 42*, 12745–12753. Available from https://doi.org/10.1039/c8nj02720j.

Veriansyah, B., Han, J. Y., Kim, S. K., Hong, S. A., Kim, Y. J., Lim, J. S., et al. (2012). Production of renewable diesel by hydroprocessing of soybean oil: Effect of catalysts. *Fuel, 94*, 578–585. Available from https://doi.org/10.1016/j.fuel.2011.10.057.

Wan Mahari, W. A., Chong, C. T., Lam, W. H., Anuar, T. N. S. T., Ma, N. L., Ibrahim, M. D., et al. (2018). Microwave co-pyrolysis of waste polyolefins and waste cooking oil: Influence of N2 atmosphere versus vacuum environment. *Energy Conversion and Management, 171*, 1292–1301. Available from https://doi.org/10.1016/j.enconman.2018.06.073.

Wang, S., Zhao, S., Uzoejinwa, B. B., Zheng, A., Wang, Q., Huang, J., et al. (2020). A state-of-the-art review on dual purpose seaweeds utilization for wastewater treatment and crude bio-oil production. *Energy Conversion and Management, 222*, 113253. Available from https://doi.org/10.1016/j.enconman.2020.113253.

Wassell, C. S., & Dittmer, T. P. (2006). Are subsidies for biodiesel economically efficient? *Energy Policy, 34*, 3993–4001. Available from https://doi.org/10.1016/j.enpol.2005.09.024.

Wellert, S., Karg, M., Imhof, H., Steppin, A., Altmann, H. J., Dolle, M., et al. (2008). Structure of biodiesel based bicontinuous microemulsions for environmentally compatible decontamination: A small angle neutron scattering and freeze fracture electron microscopy study. *Journal of Colloid and Interface Science, 325*, 250–258. Available from https://doi.org/10.1016/j.jcis.2008.05.062.

Wicaksana, A. (2016). 済無/no title.

Xu, S., Cao, B., Uzoejinwa, B. B., Odey, E. A., Wang, S., Shang, H., et al. (2020). Synergistic effects of catalytic co-pyrolysis of macroalgae with waste plastics. *Process Safety and Environmental Protection, 137*, 34–48. Available from https://doi.org/10.1016/j.psep.2020.02.001.

You, Y. D., Shie, J. L., Chang, C. Y., Huang, S. H., Pai, C. Y., Yu, Y. H., et al. (2008). Economic cost analysis of biodiesel production: Case in soybean oil. *Energy and Fuels, 22*, 182–189. Available from https://doi.org/10.1021/ef700295c.

Yusuf, N. N. A. N., Kamarudin, S. K., & Yaakub, Z. (2011). Overview on the current trends in biodiesel production. *Energy Conversion and Management, 52*, 2741–2751. Available from https://doi.org/10.1016/j.enconman.2010.12.004.

7

Waste plastic to fuel: properties, combustion characteristics and emission profile

Gaurav Kant[1], Dipesh Shikchand Patle[2], Sameer Srivastava[1] and Ashutosh Pandey[1,3,4]

[1]Department of Biotechnology, Motilal Nehru National Institute of Technology Allahabad, Prayagraj, Uttar Pradesh, India [2]Department of Chemical Engineering, Motilal Nehru National Institute of Technology Allahabad, Prayagraj, Uttar Pradesh, India [3]Institute for Water and Wastewater Technology, Durban University of Technology, Durban, South Africa [4]Faculty of Life Science and Technology, Department of Biotechnology, AKS University, Satna, Madhya Pradesh, India

7.1 Introduction

In recent decades, waste generation has skyrocketed over the globe. Global municipal solid trash generation is anticipated to increase by 70% to 3.4 billion metric tonnes by 2050. This is attributed to various causes, including population increase, urbanisation, economic growth and consumer buying patterns (Pandey, Singh, Kumar, & Srivastava, 2019). Humans produce millions of tonnes of waste yearly, which has become a serious global concern (Guerrero, Maas, & Hogland, 2013). With such massive amounts of waste being produced, authorities' need to provide proper waste treatment and disposal services has never been greater. However, approximately 20% of waste is recycled each year, with vast amounts still being dumped in landfills (Abdel-Shafy & Mansour, 2018; Guerrero et al., 2013). Waste is made up of various materials, although trash tyres have gotten a lot of attention in recent years because of their environmental effects (Guerrero et al., 2013).

Municipal solid waste (MSW) is processed to reduce the volume of waste and pollution potential of landfill sites. This can be achieved by means of material and organic recovery, recycling and

Renewable Diesel. DOI: https://doi.org/10.1016/B978-0-323-91153-5.00006-6

reuse (Millati et al., 2019). Hence, emphasis on material recovery and recycling (urban mining) is essential, as large proportions of MSW in cities comprise materials like metal and paper (Zhang, Tan, & Gersberg, 2010). In municipal solid waste management (MSWM), recycling is an important component through which waste material is converted into reusable material that is further processed for making fresh product(s) (Khan, Anjum, Raza, Ahmed Bazai, & Ihtisham, 2022; Millati et al., 2019). Recycling can be adopted for various types of wastes such as plastic, glass, paper, metal, battery, organic and electronic products. However, there are different issues related to the processes of recycling. These issues can be classified as environmental, technological, economic, legislative and social (Needhidasan, Samuel, & Chidambaram, 2014). Recycling requires high technical instrumentation or machines. They must be made in such a manner so that any kind of environmental degradation may be avoided.

Nevertheless, these types of equipment are very costly and hence they are often not feasible for the economy of underdeveloped countries (Khan et al., 2022; Lin et al., 2022). Also, recycling requires public awareness of material recovery at source to recover the maximum amount of material from MSW. This can be helped by implementing certain laws on waste producers and amendments in MSWM and handling rules. From the available literature, it can be gleaned that MSW is considered one of the resource options to manage natural resources. Moreover, it is also energy efficient to produce a new product(s) from secondary material through recycling than virgin material (Malinauskaite et al., 2017). In several developing countries, the informal sector has been involved at all levels of recycling and recovery of materials from MSW (Lin et al., 2022). Due to this practice, the recovery rate varies from 5% to 40% when the average MSW generation rate is 0.77 kg/capita/day. The waste streams of developing nations consisted of 0%–70% recyclable and 17%–80% organics (Heidbreder, Bablok, Drews, & Menzel, 2019). For example, in China, it had been estimated that recyclable wastes account for 8%–10% of the total amount of MSW on average, with an annual rate of increase of 7.1% between 1976 and 2006. This increasing rate of recyclables, specifically paper and plastics, confirms the constant decrease of organic wastes in the MSW stream. The recyclables consist of different materials with varying recovery rates. Some examples are iron and steel (85%), nonferrous metals (80%), plastics (25%), rubber (47%), paper (20%) and glass (13%). It can be expected that the material recovery and recycling industry will grow at a rapid rate in the near future around the globe as suggested in some studies. Zhang et al. (2010) suggest that 4,00,000 automobiles have been recycled in China. Nepal, on the other hand, has approximately 29% material recovery of which majorly 5%–51% consists of metal, paper, glass, plastic, rubber and leather. In Nepal although no formal system is involved in reuse and recycling, it is remarkable to notice that common people recover recyclable materials at their houses and sell them to the formal or informal sectors (Parajuly, Thapa, Cimpan, & Wenzel, 2018). In India, the proportion of recyclables in the waste stream varies from 13% to 20%. India has performed well in material recovery and recycling. For instance, the recovery of plastic waste in India is approximately 40%–80% whereas the figure for the developed countries is between 10% and 15%. In 1991 there was remarkable recovery of paper (14%), which was 37% higher than the global paper recovery rate (Brunn, 2020).

In today's scenario, recycling/recovery/management of plastic waste (PW) is a matter of concern. Industries are getting more interested in the field of plastic manufacturing, so many commodities are being manufactured with plastics. Plastics have become a crucial part of our lifestyle, and global plastic production has increased immensely during the past 50 years. It is a fact that plastics will never degrade and remain on the landscape for several years in normal environmental conditions (Earth5R, 2021). Plastic waste is harmful as its pigment contains many trace elements that are highly toxic (Heidbreder et al., 2019). As a result, environmental pollutants from synthetic plastics have been identified as a huge hassle (Heidbreder et al., 2019). Plastic solid waste (PSW) is being produced on a massive scale worldwide, and its production has crossed 150 million tonnes/year globally. It has a very broad range of applications, be it in packaging films; wrapping materials; shopping and garbage bags; fluid containers; clothing; toys; household and industrial products; and building materials. Furthermore, the recycling of virgin plastic material can be done 2 to 3 times only because, after every recycling, the strength of plastic material is reduced due to thermal degradation. Particularly, solvents having a hydrogen-donating capability take part in the thermal degradation of polymers, thereby affecting the hydrocarbon (HC) yield and distribution. It is to mention that no authentic estimate on the total generation of plastic waste is available; however, considering 70% of total plastic consumption is discarded as waste, approximately 5.6 million tonnes per annum of plastic waste is generated in the country, which is about 15,342 tonnes per day (Brunn, 2020; Heidbreder et al., 2019; Wilk & Hofbauer, 2013). This chapter, evaluate different options and processes for recycling and reusing plastic waste and also cover the various ways for converting waste plastic into energy. Furthermore, the fuel qualities, combustion, and emission characteristics of plastic waste oil are also discussed.

7.2 Plastic waste management options and processes

Plastic waste management is a way to confront plastic waste pollution and all the environmental and public health issues attributed to it. There are essentially four options for recycling plastic waste, including re-extrusion, mechanical recycling, chemical and thermal recycling, and energy recovery, as shown in Fig. 7.1. Re-extrusion of plastic waste involves the introduction of clean scrap of single types of plastics that can be re-entered into the manufacturing process. This is usually difficult and not practical as plastic scrap is usually not cleaned nor segregated based on their types at the source. It also limits the management of waste plastic to specific products. Mechanical recycling of plastic waste involves the use of one or several mechanical processes before reusing or repurposing it, such as cutting/shredding, separation of contaminants, milling, washing and drying, and extrusion. The heterogeneous mixture of PW handling is a major challenge of PW management since it requires homogenous plastic types. The segregation of plastic types is a prerequisite for mechanical recycling. The chemical and thermal recycling of plastic waste involves the use of

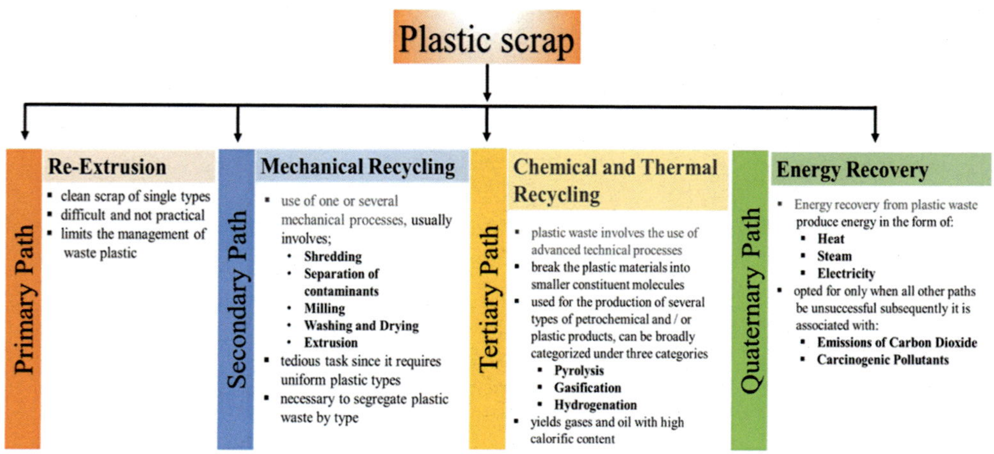

Figure 7.1 Different types of pathways plastic followed for plastic waste management.

advanced technical processes to break the plastic polymer into its monomer. The monomeric forms can then be utilised as raw materials for the production of new plastic products.

Pyrolysis, gasification and hydrogenation processes yield gases and oil with high calorific values. This path for plastic waste management is advantageous since, unlike other approaches, it can process a heterogeneous mix of plastic waste. The drawback of this pathway involves certain environmental concerns since some of the by-products can be toxic. Energy recovery from plastic waste involves the implementation of various methods to produce energy in the form of heat, steam and electricity. This path should only be opted for only when all other paths fail since it is associated with emissions of carbon dioxide and carcinogenic pollutants.

Table 7.1 presents typical examples of plastics, their coded symbols and some uses. Petroleum-based polymers, which include polypropylene (PP), polystyrene (PS), polyvinylchloride (PVC), polyethylene (PE) and polyethylene terephthalate (PETE), have been used extensively for packaging as a result of their lightweight, good mechanical performance, good barrier

Table 7.1 List of categories designated for PW management according to the plastic waste management rule (Huang et al., 2022).

Symbols	Type of plastic	Typical products
1	Polyethylene terephthalate	Package drinking bottles and some food packaging
2	High-density polyethylene	Milk bottle, detergent and vegetable oil bottles, toys, bags, tables, plastic lumber, waste bins, public places benches and bed liners of vehicles
3	Polyvinylchloride	Clear plastic food wrapping, cooking oil bottles, teething rings, blister packaging, sheathing material for computer cables, and in garden hoses, window frames, garden hoses, arbours, raised beds and trellises
4	Low-density polyethylene	Squeezable bottles, lids, toys, and plastic bags
5	Polypropylene	Disposable diapers, pails, bottle tops, yoghurt containers, potato chip bags, straws, packing tape and rope, landscaping border stripping, battery cases, brooms, bins and trays
6	Polystyrene	Rigid trays and containers, disposable eating utensils, foamed cups, plates and bowls
7	All other resins and metamaterials	Baby bottles, water cooler bottles and car parts

Source: *Adapted from Huang et al., 2022.*

properties and many other useful properties. PETE is one of the most widely used polymers in consumer goods, and it may be found in most drinking water and soda bottles, as well as certain other types of packaging. These are meant to be used only once as their reuse raises the risk of leaching and bacterial growth. PETE is difficult to disinfect, and cleaning it properly necessitates the use of hazardous chemicals. Carcinogens may seep from PE terephthalates. PETE is a recyclable plastic that is crushed and shredded into minute flakes before being recycled into new PETE bottles or spun into polyester fibre. Textiles such as fleece clothes, carpets, stuffing for mattresses and life jackets, and a few other types of similar items are made from recycled fibre. Recycling is a preferred choice for PETE plastics. PETE should move to reusable beverage containers and replace throwaway food packaging with reusable alternatives, according to PETE.

PVC is a soft, elastic-plastic that is used to produce transparent plastic food wraps, cooking oil bottles, teething rings, children's and dogs' toys, and blister packaging for a variety of consumer goods. It is typically found in computer cables, plumbing pipes and parts, and garden hoses. PVC is used to build window frames, garden hoses, arbours, raised beds and trellises since it is largely immune to sunlight and weather. PVC is known as 'poison plastic' because it includes a variety of poisons that can bleed out over time.

PVC is primarily a single-use material, and $<1\%$ of PVC is recycled. While some PCV products can be reprocessed, they should not be used in food-related applications or by children. Consider replacing plastic food wrap with reusable beeswax wraps, plastic toys with reclaimed wool plush animals and your PVC garden hose with a drinking water safe garden hose to avoid things created with PVC plastic.

Shrink wrap, dry cleaner garment bags, squeezable bottles and plastic bags used for packaging bread are all examples of low-density polyethylene (LDPE). LDPE plastic is used to make the plastic shopping bags found in most stores today. This type of plastic can also be found in apparel and furniture. LDPE is less hazardous than other plastics and is generally believed to be safe to use. However, it is not widely recycled, but this is changing in many places as more plastic recycling programmes prepare to handle this material. Plastic lumber, landscaping boards, garbage can liners and floor tiles are among the products made from recycled LDPE plastic. Recycled LDPE products are not as hard or rigid as high-density polyethylene products.

PP plastic is strong and lightweight, with excellent heat resistance. It protects against moisture, grease and chemicals. This keeps your cereal fresh and dry. Disposable diapers, pails, plastic bottle tops, margarine and yoghurt containers, potato chip bags, straws, packing tape and rope are also common uses for PP. Although PP is recyclable through some curbside recycling programmes, only about 3% of PP products are currently recycled in the United States. Landscape border striping, battery cases, brooms, bins and trays are all made from recycled PP. However, plastic is becoming more popular among recyclers. Reusing PP is thought to be safe. Check with your local garbage and recycling programme to see if they accept PP products for recycling. To reduce your PP consumption, use reusable straws instead of plastic straws, reusable water bottles and cloth diapers.

PS may seep styrene, a potential human carcinogen, into food products (especially when heated in a microwave). Chemicals found in PS have been linked to human health problems and reproductive system dysfunction. PS recycling is not widely available. Because most collection and recycling services do not accept PS, this material accounts for approximately 35% of U.S. landfill material. Whereas the technology for recycling PS is available, the recycling market is small. Customer perception has also risen, and PS is now being reused more frequently. While it is difficult to find a recycler for PS, some shipping companies are happy to receive foam packing chips for reuse. Wherever possible, PS should be avoided. Use reusable coffee cups, compostable or reusable picnic cutlery and stainless steel takeaway containers to reduce your use of PS.

Other bisphenol A (BPA), polycarbonate (PC) and LEXAN, this same seventh category was created as a catch-all for PC and 'other' plastics, so reuse and recycling protocols have not been standardised within this category. The biggest worry with the seventh category of plastics is the possibility of chemical leaching into food or drink products packaged in BPA-containing PC containers. BPA is a known xenoestrogen and endocrine disruptor. Baby bottles, Sippy cups, water cooler bottles and car parts are all made from category 7 plastics. BPA is found in PC plastic food containers, which are frequently labelled with the letters 'PC' on the bottom by the recycling label seven. Although some PC water bottles are marketed as 'nonleaching' to reduce plastic taste and odour, there is still a possibility that trace amounts of BPA will migrate from these containers, especially if heated. To replace PCs, a new generation of compostable plastics made from bio-based polymers such as corn starch is being developed. These are also included in the seventh category, which can lead to consumer confusion. These biodegradable plastics bear the initials 'polylactic acid (PLA)' near the recycling symbol on the bottom. Some may also use the term 'compostable'. Plastics in category 7 should not be reused unless they have PLA-biodegradable scripting. It is best to avoid it whenever possible, particularly when it comes to children's food. Plastic materials with recycling labels 1, 2 and 4 on the bottom are safer options because they do not contain BPA. PLA-coded plastics should be turned into compost rather than recycled because PLA-biodegradable plastics are not recyclable. The plastics industry has complied with regulatory requirements by applying the necessary codes to consumer products, but it is the responsibility of each individual to read and comprehend the codes.

7.2.1 Mechanical recycling

The recycling process is the mechanical process of recovering plastic waste through sorting, washing, drying, grinding, re-granulating and compounding. Mechanical recycling does not change the composition structure of the material, allowing for multiple reuse/recycling of polymeric materials, resulting in a circular path (Ragaert, Delva, & Van Geem, 2017). Depending on the strategy used, all end-of-life thermoplastics can be mechanically recycled with minimal quality loss.

7.2.1.1 Melt blending

Plastic deformation (extrusion) is the most common mechanical recycling method for producing re-granulated material from common waste plastics. It is inexpensive, large-scale, solvent-free and applicable to a wide range of polymers. Extruders use heat and rotating screws to induce thermal softening or plasticisation. The material is then fed through temperature-controlled barrel sections to produce fixed cross-section extruders or may occur multiple times to ensure appropriate processing (Earth5R, 2021; Ragaert et al., 2017).

7.2.1.2 Sorting

Plastics must be categorised according to their material type before they may be recycled. The first sorting of plastics might be manual sorting, which requires a lot of labour, or automated. Float-sink sorting (based on density), froth–flotation sorting, near-infrared sorting, X-ray fluorescence, laser-aided identification and marker systems are examples of sorting procedures. To achieve the necessary separation and purity, the aforementioned approaches can be applied in sequence or separately, and their application is dependent on the waste stream and available financial investment. The material is handled throughout the mechanical recycling process, and transformation procedures include shredding, grinding, sieving, density separation and other more complicated sorting operations. Mechanical recycling is the most extensively used recycling technology in Europe for treating both primary (industrial) and secondary (post-consumer) plastic waste. The process was developed to recover materials such as PE, PP and PETE. Mechanical recycling procedures often include collecting, sorting, washing, drying, grinding, compounding and pelletising. These procedures are determined by the makeup of the trash (Brunn, 2020; Ragaert et al., 2017).

7.2.1.3 Stabiliser use in plastic recycling

Both heat and light stabilisers can be used to suppress free radical reactions, including those that occur during extrusion. As a result, polymers are commonly extruded with stabilisers such as antioxidants to avoid oxidation during mechanical recycling and product usage. Primary and secondary antioxidants are the two types of antioxidants. As 'chain-breaking' antioxidants, primary antioxidant stabilisers serve as radical scavengers, forming stable peroxy radicals with oxygen and protecting chains over the polymer's lifespan. Secondary antioxidants, often based on sulphur or phosphorus, preserve polymer chains during melt processing by degrading hydroperoxide accelerants into alcohols, functioning as 'preventative' antioxidants. Stabilisers can also absorb and diffuse light energy, protecting chains from ultraviolet (UV)-based deterioration. Polymers are often only stabilised during their initial life cycle since the packaging is supposed to be disposable, underlining the necessity for polymer design to include end-of-life recycling considerations (Phaniendra, Jestadi, & Periyasamy, 2015; Shao, Chu, Lu, & Kang, 2008).

7.2.1.4 Polymer blends in recycling

Mixed polymer streams, however, common in mechanical recycling, generally result in weaker products due to polymer phase immiscibility. These melt incompatibilities cause fracture sites in the extruded material (Dorigato, 2021). The Flory-Huggins hypothesis predicts that when the Gibbs free energy of mixing for two polymers is great enough, the blending will be favoured, and phase separation will occur. Demixed polymers can take on a variety of phase separation morphologies, such as spheres, cylinders, lamellae and continuous segments. Due to inadequate stress and strain transmission across phase boundaries, phase separation in blends leads to poor mechanical characteristics. Increase the frequency of contacts between phases to promote stress transmission across borders. Improved polymer blend compatibility would boost the value of recycling by improving process capability, plant flexibility, product tailoring and mixed recyclate performance. Compatibilisers avoid extreme remixing of two polymer phases and promote blend stability whenever the phase separation is >10% by lowering interfacial tension between the two phases to promote dispersion, stable phase morphology and enhanced phase adhesion. Improving adhesion allows for optimal stress and strain transmission within the mix. Physical deterioration happens in nature as a result of pressure, humidity and sun heat. Based on their chemical constituents, the bonds between the HC monomers in plastics composed of HC chain polymers derived from petroleum refining are tough and durable.

7.2.2 Chemical recycling

Plastic is a wide range of synthetic or semi-synthetic materials that are used in a huge and growing range of applications. Chemical recycling pathways provide the finest possibilities for wide-scale application since they enable the generation of fuels, chemicals and syngas/hydrogen from waste plastics. The main chemical valorisation routes of waste plastics are shown in Fig. 7.2. These include processes such as pyrolysis, gasification,

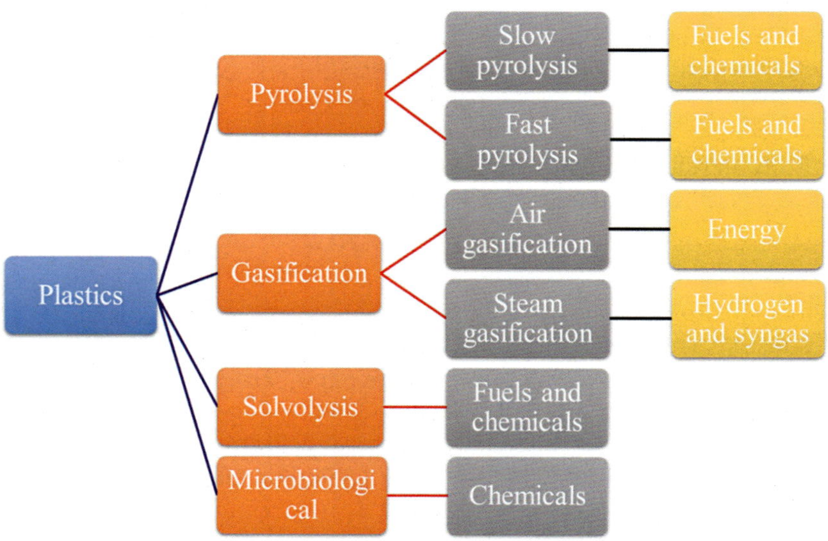

Figure 7.2 Routes for the chemical valorisation of waste plastics.

solvolysis and microbiological, which breakdown plastic waste into monomers, the chemical building blocks used to create new plastic materials (Lopez et al., 2018). However, certain types of plastics, such as single-use plastics and multi-layer laminated plastics, are difficult to recycle.

7.2.2.1 Pyrolysis

Pyrolysis is the thermal breakdown of organic molecules in the absence of oxygen, producing gases, liquids and solids. At high temperatures (300°C–600°C), long-chain HCs are fractured into tiny-chain HCs by thermal degradation. There are three primary processes involved in the pyrolytic reaction: commencement, propagation and termination. The lengthy carbon chains in plastics are thermally broken down into usable components that may be used as fuels or chemical sources during pyrolysis. A liquid product, a gaseous product and a solid product are typically created. Plastic waste from various types of synthetic plastic is shredded and combined with a catalyst in certain quantities and ratios in the pyrolysis reactor, as shown in Table 7.2. The yield and composition of pyrolysis products vary owing to numerous parameters such as feedstock type, reactor system, gas residence time, heating rate, contact time, temperature, pressure ranges, the action of catalysts (if present) and the presence of hydrogen gas or hydrogen giving chemicals. The reactor is made of stainless steel and has a permanent bed. The plastic is heated at a rate of 15°C/min to a maximum temperature of 550°C. The pyrolysis process produces oil and gas vapour. These products are condensed to generate fuel oil, heavy oil and light HC gas. Because HCs are saved and reused throughout the combustion process, this system produces its

Table 7.2 Comparison between fast and slow pyrolysis.

S. no.	Characteristics	Fast pyrolysis	Slow pyrolysis
1	Reactor type	Fluidised and kiln reactor	Fixed bed
2	Main product	Liquid phase	Solid phase
3	Particle size	Small	Relatively big
4	Temperature	Higher	Lower
5	Residence time	Few seconds	Minutes to hours
6	Heating rate	High	Low
7	Degradation rate	Fast	Slow

own energy. Biofuel oil is delivered to refineries and transformed into gasoline and diesel using the heterogeneous catalyst Zeolite Socony Mobil-5. Temperature, residence time, reactor type, pressure, type and rate of fluidising gas, heating rate, type of catalyst and feedstock all have a significant impact on the properties and yield of the products. Pyrolysis is further subdivided into slow and rapid pyrolysis. Table 7.2 compares slow and quick pyrolysis in terms of the factors listed above. Slow pyrolysis reduces the processing efficiency of the pyrolysis process; hence, researchers have focused their efforts on quick pyrolysis (Tan et al., 2021). Another type of pyrolysis technique is microwave-assisted pyrolysis. Unlike typical heating, which transfers heat from the feedstock's surface to its core, microwave-assisted pyrolysis carries energy throughout the feedstock's volume (Singh et al., 2022). Furthermore, because the feedstock is heated instantly in microwave-assisted pyrolysis, there is no need for feedstock shredding (Singh et al., 2022). It also improves heating quality, which leads to the manufacture of value-added products. It is appropriate for large-scale commercialisation due to its low cost of production, the versatility of the procedure, and the convenience with which equipment may be transported. The pyrolysis of plastic waste has two advantages: it reduces landfill dumping and recovers valuable HCs (Aishwarya & Sindhu, 2016; Singh et al., 2022).

7.2.2.2 Solvolysis

Solvolysis is the employment of solvents, heat and pressure to carry out the reverse reaction for the production of polymers. Depending on the chemical agent employed for polymer chain scission, different monomers or oligomers are generated depending on the solvent (methanol, ethylene glycol or water) (Shi et al., 2016).

7.2.2.3 Gasification (case studies)

Gasification is the partial oxidation of organic waste at temperatures lower than combustion. As a beneficial by-product of this process, syngas is produced. Gasification of waste plastics produces a stream mostly composed of H_2, CO, CO_2, CH_4 and N_2. Thus the practicality of creating energy, energy carriers (such as H_2) and chemicals (fuels, DME, methanol and so on) from syngas generated is at the heart of gasification operations (Sansaniwal, Pal, Rosen, & Tyagi, 2017). The increased flexibility to jointly valorise plastics of varied compositions or mixes, or plastics combined with other feedstocks, is a significant benefit of gasification over pyrolysis. The composition of the gas generated, and hence its uses, is determined by the gasifying agent utilised. The air gasification of waste plastics produces syngas with an estimated calorific value of 6–8 MJ m^3 (Arena & Di Gregorio, 2014), with the primary goal being energy generation. Nonetheless, steam gasification allows for the production of N_2-free syngas with a heating value of more than 15 MJ m^3 and a composition appropriate for synthesis applications (Erkiaga, Lopez, Amutio, Bilbao, & Olazar, 2013; Wilk & Hofbauer, 2013). The fundamental problem of waste plastics gasification is the high tar concentration in the gas output, which is often greater than that observed for biomass gasification (Erkiaga et al., 2013; Lopez et al., 2018). The process of tar generation, particularly its evolution in the reaction environment in plastic waste gasification, differs significantly from that of coal and biomass, owing to the differing content of the volatiles generated. In the case of plastic trash, the method of tar production and development is determined by the nature of the plastic waste (Fig. 7.3). In reality, the direct synthesis of primary aromatic tar happens exclusively during the decomposition of polymers containing aromatic rings in their structure, primarily PS and PET. Furthermore, the volatiles or main tar generated from the decomposition of polyolefins are alkanes and alkenes with different chain lengths. Plastic gasification aims for maximal conversion to agas products or syngas, with tar and char as the principal unwanted by-products.

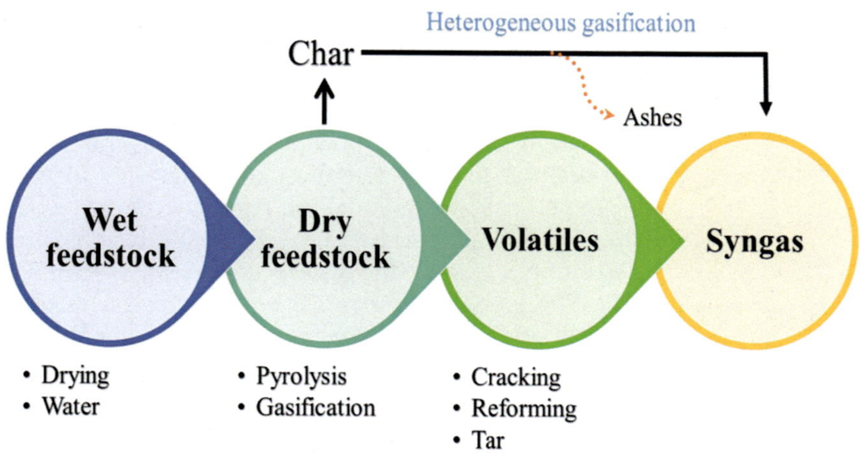

Figure 7.3 Crucial steps involved in plastics gasification. Adopted from Lopez, G., Artetxe, M., Amutio, M., Alvarez, J., Bilbao, J., & Olazar, M. (2018). Recent advances in the gasification of waste plastics: A critical overview. *Renewable and Sustainable Energy Reviews, 82*, 576–596. https://doi.org/10.1016/j.rser.2017.09.032.

Gasification consists of multiple processes and complicated chemical reactions, but they may be described as follows: drying, pyrolysis, cracking and reforming events in the gas phase, and heterogeneous char gasification (Fig. 7.3).

The plastics degradation process is important in the decomposition of the volatiles generated and hence in the following cracking and reforming events that occur in the gaseous phase. The interest in plastic waste gasification research is new, and the number of studies is expectedly small. Several investigations on co-gasification with coal and biomass have been conducted. Direct air gasification is the best researched of these, resulting in a gas product with a very low heating value due to the diluting action of N_2. The most explored and proven technique is air gasification, which attempts to create syngas for energy usage. Despite its increased H_2 content and heating value, gasified steam gas has significant challenges, such as the process's energy needs and the tar level in the syngas. Furthermore, the co-gasification of plastics with coal and biomass looks to be a viable valorisation alternative due to the positive influence on process performance and greater process flexibility. Another viable possibility is pyrolysis and in-line reforming, which produce syngas with a high hydrogen concentration and no tar (Lopez et al., 2018). As a result, this gas is mostly employed in the generation of energy (Arena & Di Gregorio, 2014; Bernocco, Bosio, & Arato, 2013).

7.2.2.4 Microbiological

The polymer is broken into smaller fragments, such as monomers and other industrially valuable compounds, in this context. This degradation is carried out by bacteria or enzymes that have been released. It is a really uncommon form of breakup. Microbial degradation of plastic polymers involves the phases of bio-deterioration, bio-fragmentation, assimilation and mineralisation. The superficial degradation of plastic surfaces, which modifies their physical, chemical and mechanical characteristics, is known as bio-deterioration (Ghosh, Pal, & Ray, 2013; Pathak & Navneet, 2017). The growth of microbial biofilms on and inside the plastic substrate adds to the degradation process. It produces considerable physicochemical deterioration, the creation of which is reliant not only on the structural properties of the plastic polymers but also on the environmental circumstances. Following bio-deterioration comes bio-fragmentation, which includes the catalytic activity of microbial enzymes on plastic polymers (Pathak & Navneet, 2017). Bacteria that can degrade plastic polymers often contain oxygenases (monooxygenases and dioxygenases), which add oxygen molecules into a long chain of carbon, generating less resistant and less hazardous alcohol and peroxyl compounds. Furthermore, lipases and esterases catalyse carboxylic group transformations, while endopeptidases catalyse amide group transformations. Plastic polymers have been broken down by *Brevibacillus borstelensis, Rhodococcus ruber, Pseudomonas chlororaphis, Pseudomonas putida* AJ, *Thermomonospora fusca, Alcaligenes faecalis, Clostridium* sp. and others (Vivi, Martins-Franchetti, & Attili-Angelis, 2019). Mineralisation and absorption are the ultimate phases of microbial degradation of polymeric polymers. Plastic monomers generated by bio-fragmentation must pass through the microbial cell membrane. The plastic monomers were oxidised by catabolic processes inside the cells, and the energy generated was utilised to produce new cell biomass. Assimilation is the insertion of atoms into the microbial cell that allows degradation to take place. Metabolite breakdown produces oxidised chemicals such as CO_2, N_2, CH_4 and H_2O (Jaiswal, Sharma, & Shukla, 2020).

7.3 Fuel properties, combustion and emission characteristics

Crude oil imports have faced difficulties in recent years as a result of rising crude oil prices, a volatile political situation in the Middle East and the unpredictability of the global oil market (The World Bank, n.d.; U.S. Energy Information Administration, n.d.). The use of fossil fuels also has negative environmental consequences for land, water and air. The use of fossil fuels has the potential to significantly alter the global climate (Pandey, Srivastava, & Kumar, 2019a). The combustion of fossil fuels creates energy and chemicals, which results in the release of numerous greenhouse gases such as carbon dioxide, nitrogen oxides (NO_x) and other hazardous volatile compounds into the atmosphere (Pandey, Srivastava, & Kumar, 2019b). As a result, throughout the last few decades, the selection of the most environmentally friendly fuel has been a key priority. Municipalities have collected and disposed of a large amount of plastic waste as a result of increasing urbanisation and the expansion of the global economy. Because of the dramatic increase in the disposal of plastic garbage, its management is critical to the preservation of the environment (Abdel-Shafy & Mansour, 2018; Wan Mahari et al., 2018). Waste plastics (WP) are often converted by pyrolysis, which converts the plastic polymers into their fundamental monomers or HCs; this method is referred to as tertiary (or chemical) recycling, as discussed earlier in this chapter (Papari, Bamdad, & Berruti, 2021).

Despite the volume of work and efforts on the conversion of WP into fuel, only a handful of them undertakes a comprehensive study on the combustion of such fuels in high-speed diesel engines. A group of researchers evaluated various waste plastic oil and heavy oil (type A) mixes on a single-cylinder diesel engine with a low-pressure (approximately 200 bar) direct injection system. They discovered that the increased viscosity of WP blends, when compared to heavy oil, enhances injection duration (Mani, Nagarajan, & Sampath, 2011). As a result, peak pressures are lower with the WP blend, and in-cylinder pressure rises more steadily. In terms of engine performance, they observed that the thermal efficiency of polymer fuels is nearly comparable to that of heavy oil, signifying that thermal recycling of plastic waste is viable for engine use. Mani et al. performed a thorough test on a single-cylinder Kikloskar Direct Injection research engine with a total displacement of around 600 cc and a rated power of 4.4 kW at 1500 rpm (Mani et al., 2011). They used both pure waste plastic oil (WPO) and a blend of WPO and diesel oil to feed the engine without any alterations. They discovered that the maximum heat release rate with WPO is greater than with diesel due to a longer ignition delay in the WPO engine, which results in a larger in-cylinder pressure peak (Mani et al., 2011). Smoke emissions for WPO, on the other hand, were reduced considerably (up to 40%–50%), and even the tested blends produced lower soot emissions; nonetheless, NOx exhaust concentrations were found to be greater (Pan et al., 2019). In a similar kind of study, Mani et al. showed that by modifying the injection time, the engine performance and emissions of a WPO-powered engine might be improved even further. Güngör et al. put a plastic oil made from waste PE through its paces on a Mitsubishi 4D34-2A engine rated at 89 kW at 3200 rpm and observed that the blends' fuel attributes are equivalent to those of diesel fuel as defined by the EN-590 Diesel Fuel Standard. Moreover, using WPO mixes enhanced the engine's power output somewhat while lowering carbon monoxide (CO) emissions by 20.6%. However, there was an increase in NO_x (Güngör, Serin, Özcanlı, Serin, & Aydın, 2015). Kumar et al. evaluated the performance of WPO blends on a Comet VCT-10 engine and found that, due to the lower heating value of their WPO, braking thermal efficiency at full load is lower than that of diesel fuel (Kumar, Mishra, Singh, & Kumar, 2016). CO, NO_x, unburned HC and smoke were emitted by the combustion of distilled WPO and diesel fuel, respectively. The NO_x emissions decreased as the engine operating load increased for all test fuels (Güngör et al., 2015; Kumar et al., 2016; Rinaldini et al., 2016).

When an engine is driven at a high engine load, a lesser oxygen availability might be taken into consideration for a reduction in NO_x. As compared with diesel fuel, the NO_x emissions of WPO were higher than those of diesel fuel (Güngör et al., 2015; Rajesh Kumar, Saravanan, & Rajaram, 2019) (Table 7.3).

This was due to the longer ignition delay owing to the long-chain carbon compounds in the WPO. Furthermore, a larger compression ratio might improve the combustion process, but potentially at the expense of increased NO_x emissions. CO is a by-product of incomplete combustion, which is enhanced by a lack of oxygen, in-cylinder temperature, and combustion residence time (Mani et al., 2011). For all test fuels, increasing engine running loads resulted in lower CO emissions. The drop in CO emissions might be attributable to an improvement in combustion efficiency as the engine's operational load rises (Kumar et al., 2016). When comparing the test fuels, the burning of WPO resulted in greater CO emissions. The increased CO emissions from the usage of WPO may be explained by the lower cetane index of WPO, which tended to delay the combustion state,

Table 7.3 Physicochemical properties of waste plastic oil (WPO), crude waste plastic oil (CWPO) and diesel fuel.

Properties	WPO		CWPO	Diesel
	A	B		
Ash content (wt.%)	0.036	—	—	—
Calorific value (kcal/kg)	9829.35	45.24	37.72	45.56
Carbon residue (wt.%)	0.5	—	—	—
Density at 40°C (g/cc)	0.7477	0.823	0.899	0.834
Fire point (°C)	20	—	—	—
Flash point (°C)	15	54	35	66
Pour point (°C)	<−15	—	—	—
Sulphur content (wt.%)	0.246	—	—	—
Viscosity at 40°C (cSt)	1.980	3.11	1.66	3.44
Surface tension (mN/m)	—	27.85	27.77	
Specific gravity at 15.6°C	—	0.824	0.900	0.835
Cetane index	—	46.7		
Distillation temperature (°C)				
10% Recovered (°C)	—	232	112	228
50% Recovered (°C)	—	276	164	290
90% Recovered (°C)	—	324	252	348

resulting in a reduction in the residence duration of combustion. Furthermore, a larger proportion of long-chain carbon compounds (C_{12}–C_{20}) in WPO tended to make combustion more challenging, resulting in an increase in CO emissions (Rinaldini et al., 2016). WPO CO emissions tended to decrease as the compression ratio increased, and the engine was operated at full power. Increased compression ratios resulted in higher combustion efficiency due to higher charged air temperature, which resulted in improved air–fuel mixes and faster fuel vaporisation, resulting in complete burning (Kumar et al., 2016).

Smoke emission is a sign of inefficient combustion caused by an overly rich air–fuel combination. When a rich mixture is burned in the cylinder, smoke emissions rise as engine operating loads rise. The burning of WPO resulted in fewer smoke emissions when compared to diesel fuel (Padmanabhan et al., 2022). The diesel fuel utilised in this study is regular grade and contains 7% biodiesel, according to the researchers. The presence of oxygen in biodiesel is clearly linked to a reduction in smoke emissions in general. Smoke emissions from regular diesel fuel are projected to be lower than those from WPO. The lower smoke emissions could be attributed to WPO's lower viscosity and lower distillation temperature, which can promote fuel atomisation and hence improve the combustion process (Mani et al., 2011; Rajesh Kumar et al., 2019). When the WPO was run at low and medium loads with a high compression ratio, these two characteristics were more likely to contribute to the reduction in smoke emissions. When the engine was run at maximum load, however, WPO combustion with an excessively rich air–fuel ratio tended to increase smoke emissions as the compression ratio increased. This means that at the highest engine load and compression ratio, the effect of decreased viscosity and distillation temperature cannot reduce emissions (Güngör et al., 2015; Kumar et al., 2016; Wang, Liu, & Reitz, 2017).

7.4 Conclusions and recommendations

The ideal path to achieving a sustainable society is to replace synthetic plastics. A plastic-free world, on the other hand, is now a dream, and significant effort must be committed in the pursuit of a major change in end-of-life plastic waste treatment and management. Plastic degradation in nature is classified into three categories: physical, biological and chemical processes. Nature's physical degradation processes occur as a result of pressure, humidity, solar heat and UV. Based on their chemical composition, the bonds between the HC monomers in plastics formed of HC chain polymers obtained from petroleum refining are so strong that the degradation process at ambient temperature is difficult. As a result, it is highly difficult to be biologically destroyed by enzymes and bacteria, and the degradation process takes a very long time. Several solutions have thus been created to carry out efficient recycling of plastic trash, including physical and chemical procedures. Several possibilities have indeed been created to carry out an efficient process of recycling plastic waste, including physical and chemical treatments.

Physical techniques are frequently referred to as the 'three Rs', which stand for reduction, reuse and recycling. This strategy is ineffective since plastic waste returns to the system and ultimately becomes waste plastic. Furthermore, the separation procedure necessitates substantial labour expenses, decreasing the business's sustainability. However, the chemical process of pyrolysis is a more efficient and cost-effective form of recycling since it can recover the energy inherent in plastic. Although investigators have focused on chemical treatments in the past few years, mechanical recycling is still the most developed and commercial technology. The absence of appropriate infrastructures and technology is impeding the commercialisation of chemical upcycling as well as the replacement of present materials with more sustainable polymers. Prospective remedies will primarily concentrate on the improvement of environmentally friendly materials, totally recyclable polymers and depolymerisation pathways that maximise the life cycle of plastics.

Acknowledgement

G.K would like to acknowledge the Department of Biotechnology (DBT), New Delhi, India for Junior Research Fellowship (JRF) and Senior Research Fellowship (SRF) during PhD programme.

References

Abdel-Shafy, H. I., & Mansour, M. S. M. (2018). Solid waste issue: Sources, composition, disposal, recycling, and valorisation. *Egyptian Journal of Petroleum, 27,* 1275–1290. Available from https://doi.org/10.1016/j.ejpe.2018.07.003.

Aishwarya, K. N., & Sindhu, N. (2016). Microwave assisted pyrolysis of plastic waste. *Procedia Technology, 25,* 990–997. Available from https://doi.org/10.1016/j.protcy.2016.08.197.

Arena, U., & Di Gregorio, F. (2014). Energy generation by air gasification of two industrial plastic wastes in a pilot scale fluidised bed reactor. *Energy, 68*, 735–743. Available from https://doi.org/10.1016/j.energy.2014.01.084.

Bernocco, D., Bosio, B., & Arato, E. (2013). Feasibility study of a spouted bed gasification plant. *Chemical Engineering Research and Design., 91*, 843–855. Available from https://doi.org/10.1016/j.cherd.2012.09.012.

Brunn, M. (2020). Waste management crisis in India. *Recycling Magazine.* <https://www.recycling-magazine.com/2020/05/06/waste-management-crisis-in-india/> Accessed 19.05.22.

Dorigato, A. (2021). Recycling of polymer blends. *Advanced Industrial and Engineering Polymer Research, 4*, 53–69. Available from https://doi.org/10.1016/j.aiepr.2021.02.005.

Earth5R. (2021). *Recycling as a solution to better plastic waste management at Kolkata, India.* <https://earth5r.org/recycling-as-a-solution-to-the-plastic-waste-management-at-kolkata-india/> Accessed 19.05.22.

Erkiaga, A., Lopez, G., Amutio, M., Bilbao, J., & Olazar, M. (2013). Syngas from steam gasification of polyethylene in a conical spouted bed reactor. *Fuel, 109*, 461–469. Available from https://doi.org/10.1016/j.fuel.2013.03.022.

Ghosh, S. K., Pal, S., & Ray, S. (2013). Study of microbes having potentiality for biodegradation of plastics. *Environmental Science and Pollution Research., 20*, 4339–4355. Available from https://doi.org/10.1007/s11356-013-1706-x.

Guerrero, L. A., Maas, G., & Hogland, W. (2013). Solid waste management challenges for cities in developing countries. *Waste Management (New York, NY), 33*, 220–232. Available from https://doi.org/10.1016/j.wasman.2012.09.008.

Güngör, C., Serin, H., Özcanlı, M., Serin, S., & Aydın, K. (2015). Engine performance and emission characteristics of plastic oil produced from waste polyethylene and its blends with diesel fuel. *International Journal of Green Energy, 12*, 98–105. Available from https://doi.org/10.1080/15435075.2014.893873.

Heidbreder, L. M., Bablok, I., Drews, S., & Menzel, C. (2019). Tackling the plastic problem: A review on perceptions, behaviors, and interventions. *The Science of the Total Environment, 668*, 1077–1093. Available from https://doi.org/10.1016/j.scitotenv.2019.02.437.

Huang, S., Wang, H., Ahmad, W., Ahmad, A., Ivanovich Vatin, N., Mohamed, A. M., … Mehmood, I. (2022). Plastic waste management strategies and their environmental aspects: A scientometric analysis and comprehensive review.. *International Journal of Environmental Research and Public Health, 19*(8). Available from https://doi.org/10.3390/ijerph19084556.

Jaiswal, S., Sharma, B., & Shukla, P. (2020). Integrated approaches in microbial degradation of plastics. *Environmental Technology & Innovation, 17*, 100567. Available from https://doi.org/10.1016/j.eti.2019.100567.

Khan, S., Anjum, R., Raza, S. T., Ahmed Bazai, N., & Ihtisham, M. (2022). Technologies for municipal solid waste management: Current status, challenges, and future perspectives. *Chemosphere, 288*, 132403. Available from https://doi.org/10.1016/j.chemosphere.2021.132403.

Kumar, R., Mishra, M. K., Singh, S. K., & Kumar, A. (2016). Experimental evaluation of waste plastic oil and its blends on a single cylinder diesel engine. *Journal of Mechanical Science and Technology, 30*, 4781–4789. Available from https://doi.org/10.1007/s12206-016-0950-7.

Lin, K., Zhao, Y., Kuo, J.-H., Deng, H., Cui, F., Zhang, Z., … Wang, T. (2022). Toward smarter management and recovery of municipal solid waste: A critical review on deep learning approaches. *Journal of Cleaner Production, 346*, 130943. Available from https://doi.org/10.1016/j.jclepro.2022.130943.

Lopez, G., Artetxe, M., Amutio, M., Alvarez, J., Bilbao, J., & Olazar, M. (2018). Recent advances in the gasification of waste plastics: A critical overview. *Renewable and Sustainable Energy Reviews, 82*, 576–596. Available from https://doi.org/10.1016/j.rser.2017.09.032.

Malinauskaite, J., Jouhara, H., Czajczyńska, D., Stanchev, P., Katsou, E., Rostkowski, P., ... Spencer, N. (2017). Municipal solid waste management and waste-to-energy in the context of a circular economy and energy recycling in Europe. *Energy, 141*, 2013–2044. Available from https://doi.org/10.1016/j.energy.2017.11.128.

Mani, M., Nagarajan, G., & Sampath, S. (2011). Characterisation and effect of using waste plastic oil and diesel fuel blends in compression ignition engine. *Energy, 36*, 212–219. Available from https://doi.org/10.1016/j.energy.2010.10.049.

Millati, R., Cahyono, R. B., Ariyanto, T., Azzahrani, I. N., Putri, R. U., & Taherzadeh, M. J. (2019). *Agricultural, industrial, municipal, and forest wastes. Sustainable resource recovery and zero waste approaches* (pp. 1–22). Elsevier. Available from https://doi.org/10.1016/B978-0-444-64200-4.00001-3.

Needhidasan, S., Samuel, M., & Chidambaram, R. (2014). Electronic waste – an emerging threat to the environment of urban India. *Journal of Environmental Health Science and Engineering, 12*, 36. Available from https://doi.org/10.1186/2052-336X-12-36.

Padmanabhan, S., Giridharan, K., Stalin, B., Kumaran, S., Kavimani, V., Nagaprasad, N., ... Krishnaraj, R. (2022). Energy recovery of waste plastics into diesel fuel with ethanol and ethoxy ethyl acetate additives on circular economy strategy. *Scientific Reports, 12*, 5330. Available from https://doi.org/10.1038/s41598-022-09148-2.

Pan, M., Zheng, Z., Huang, R., Zhou, X., Huang, H., Pan, J., & Chen, Z. (2019). Reduction in PM and NO_X of a diesel engine integrated with n-octanol fuel addition and exhaust gas recirculation. *Energy, 187*, 115946. Available from https://doi.org/10.1016/j.energy.2019.115946.

Pandey, A., Singh, M. P., Kumar, S., & Srivastava, S. (2019). Phycoremediation of Persistent Organic Pollutants from Wastewater: Retrospect and Prospects. In S. K. Gupta, & F. Bux (Eds.), *Application of Microalgae in Wastewater Treatment*. Cham: Springer. Available from https://doi.org/10.1007/978-3-030-13913-1_11.

Pandey, A., Srivastava, S., & Kumar, S. (2019a). Isolation, screening and comprehensive characterisation of candidate microalgae for biofuel feedstock production and dairy effluent treatment: A sustainable approach. *Bioresource Technology, 293*, 121998. Available from https://doi.org/10.1016/j.biortech.2019.121998.

Pandey, A., Srivastava, S., & Kumar, S. (2019b). Sequential optimisation of essential nutrients addition in simulated dairy effluent for improved Scenedesmus sp ASK22 growth, lipid production and nutrients removal. *Biomass Bioenergy, 128*, 105319. Available from https://doi.org/10.1016/j.biombioe.2019.105319.

Papari, S., Bamdad, H., & Berruti, F. (2021). Pyrolytic conversion of plastic waste to value-added products and fuels: A review. *Materials, 14*, 2586. Available from https://doi.org/10.3390/ma14102586.

Parajuly, K., Thapa, K. B., Cimpan, C., & Wenzel, H. (2018). Electronic waste and informal recycling in Kathmandu, Nepal: Challenges and opportunities.

Journal of Material Cycles and Waste Management, *20*, 656–666. Available from https://doi.org/10.1007/s10163-017-0610-8.

Pathak, V. M., & Navneet. (2017). Review on the current status of polymer degradation: A microbial approach. *Bioresources and Bioprocessing.*, *4*, 15. Available from https://doi.org/10.1186/s40643-017-0145-9.

Phaniendra, A., Jestadi, D. B., & Periyasamy, L. (2015). Free radicals: Properties, sources, targets, and their implication in various diseases. *Indian Journal of Clinical Biochemistry*, *30*, 11–26. Available from https://doi.org/10.1007/s12291-014-0446-0.

Ragaert, K., Delva, L., & Van Geem, K. (2017). Mechanical and chemical recycling of solid plastic waste. *Waste Management (New York, NY)*, *69*, 24–58. Available from https://doi.org/10.1016/j.wasman.2017.07.044.

Rajesh Kumar, B., Saravanan, S., & Rajaram, K. (2019). Combined effect of oxygenates and injection timing for low emissions and high performance in a diesel engine using multi-response optimisation. *Alexandria Engineering Journal*, *58*, 625–636. Available from https://doi.org/10.1016/j.aej.2019.03.009.

Rinaldini, C. A., Mattarelli, E., Savioli, T., Cantore, G., Garbero, M., & Bologna, A. (2016). Performance, emission and combustion characteristics of a IDI engine running on waste plastic oil. *Fuel*, *183*, 292–303. Available from https://doi.org/10.1016/j.fuel.2016.06.015.

Sansaniwal, S. K., Pal, K., Rosen, M. A., & Tyagi, S. K. (2017). Recent advances in the development of biomass gasification technology: A comprehensive review. *Renewable and Sustainable Energy Reviews*, *72*, 363–384. Available from https://doi.org/10.1016/j.rser.2017.01.038.

Shao, H.-B., Chu, L.-Y., Lu, Z.-H., & Kang, C.-M. (2008). Primary antioxidant free radical scavenging and redox signaling pathways in higher plant cells. *International Journal of Biological Sciences*, 8–14. Available from https://doi.org/10.7150/ijbs.4.8.

Shi, Y., Xia, X., Li, J., Wang, J., Zhao, T., Yang, H., ... Jiang, X. (2016). Solvolysis kinetics of three components of biomass using polyhydric alcohols as solvents. *Bioresource Technology*, *221*, 102–110. Available from https://doi.org/10.1016/j.biortech.2016.09.008.

Singh, A., Kushwaha, A., Sen, S., Goswami, S., Katiyar, S., Kumar, A., ... Hussain, C. M. (2022). *Recent advancement in microwave-assisted pyrolysis for biooil production. Waste-to-energy approaches towards zero waste* (pp. 197–219). Elsevier. Available from https://doi.org/10.1016/B978-0-323-85387-3.00014-8.

Tan, H., Lee, C. T., Ong, P. Y., Wong, K. Y., Bong, C. P. C., Li, C., & Gao, Y. (2021). A review on the comparison between slow pyrolysis and fast pyrolysis on the quality of lignocellulosic and lignin-based biochar. *IOP Conference Series: Materials Science and Engineering*, *1051*, 012075. Available from https://doi.org/10.1088/1757-899X/1051/1/012075.

The World Bank. (n.d.). *Solid waste management*. <https://www.worldbank.org/en/topic/urbandevelopment/brief/solid-waste-management> Accessed 18.05.22.

U.S. Energy Information Administration. (n.d.). *Oil and petroleum products explained: Oil imports and exports*. <https://www.eia.gov/energyexplained/oil-and-petroleum-products/imports-and-exports.php> Accessed 18.05.22.

Vivi, V. K., Martins-Franchetti, S. M., & Attili-Angelis, D. (2019). Biodegradation of PCL and PVC: Chaetomium globosum (ATCC 16021) activity. *Folia Microbiologica*, *64*, 1–7. Available from https://doi.org/10.1007/s12223-018-0621-4.

Wan Mahari, W. A., Chong, C. T., Lam, W. H., Anuar, T. N. S. T., Ma, N. L., Ibrahim, M. D., & Lam, S. S. (2018). Microwave co-pyrolysis of waste polyolefins and waste cooking oil: Influence of N2 atmosphere versus vacuum environment. *Energy Conversion and Management, 171*, 1292–1301. Available from https://doi.org/10.1016/j.enconman.2018.06.073.

Wang, Z., Liu, H., & Reitz, R. D. (2017). Knocking combustion in spark-ignition engines. *Progress in Energy and Combustion Science, 61*, 78–112. Available from https://doi.org/10.1016/j.pecs.2017.03.004.

Wilk, V., & Hofbauer, H. (2013). Conversion of mixed plastic wastes in a dual fluidised bed steam gasifier. *Fuel, 107*, 787–799. Available from https://doi.org/10.1016/j.fuel.2013.01.068.

Zhang, D. Q., Tan, S. K., & Gersberg, R. M. (2010). Municipal solid waste management in China: Status, problems and challenges. *Journal of Environmental Management, 91*, 1623–1633. Available from https://doi.org/10.1016/j.jenvman.2010.03.012.

Further reading

Das, A. K., & Panda, A. K. (2023). Improvement of performance and emission characteristic of waste plastic pyrolytic oil blended diesel fuel in variable compression diesel engine using graphene nano additive. *Energy & Environment, 34*(1), 212–229. Available from https://doi.org/10.1177/0958305X211060405.

Padmanabhan, S., Vinod Kumar, T., Giridharan, K., Stalin, B., Nagaprasad, N., Leta Tesfaye Jule., & Krishnaraj, R. (2022). An analysis of environment effect on ethanol blends with plastic fuel and blend optimization using a full factorial design. *Scientific Reports, 12*(1), 21719.

Singh, R. K., Biswajit Ruj., Sadhukhan, A. K., & Gupta, P. (2020). Waste plastic to pyrolytic oil and its utilization in CI engine: Performance analysis and combustion characteristics. *Fuel, 262*, 116539.

Singh, T. S., Rajak, U., Dasore, A., Muthukumar, M., & Verma, T. N. (2021). Performance and ecological parameters of a diesel engine fueled with diesel and plastic pyrolyzed oil (PPO) at variable working parameters. *Environmental Technology & Innovation, 22*, 101491.

Wathakit, K., Sukjit, E., Kaewbuddee, C., Maithomklang, S., Klinkaew, N., Liplap, P., ... Srisertpol, J. (2021). Characterization and impact of waste plastic oil in a variable compression ratio diesel engine. *Energies, 14*(8), 2230.

Index

Note: Page numbers followed by "*f*" and "*t*" refer to figures and tables, respectively.

R